ISSN 0021-8359

Heft 5 · 1978 · Band 19

Journal für Hirnforschung

Internationales Journal für Neurobiologie

Begründet von Cécile und Oskar Vogt

Akademie-Verlag · Berlin

EVP 25,- M · 32105

Begründet von Cécile und Oskar Vogt
Unter Mitwirkung des Cécile und Oskar Vogt Instituts für Hirnforschung in Düsseldorf und der Arbeitsgemeinschaft für vergleichende Neuroanatomie der Fédération mondiale de Neurologie (World Federation of Neurology)

Im JOURNAL FÜR HIRNFORSCHUNG werden Arbeiten aus dem Gesamtgebiet der normalen Morphologie (Anatomie, Histologie, Cytologie, Elektronenmikroskopie, Histochemie) und der Entwicklungsgeschichte des Nervensystems unter Einschluß experimentell-anatomischer Arbeiten veröffentlicht. Es werden auch Arbeiten multidisziplinären Charakters aufgenommen, sofern sie morphologische Ergebnisse beinhalten, die mit neuromorphologischen und neurophysiologischen oder neuropharmakologischen bzw. neurochemischen Methoden gewonnen wurden und einen Erkenntnisgewinn hinsichtlich der Wechselwirkung zwischen Struktur und Funktion beinhalten. Neuropathologische Arbeiten werden nur angenommen, wenn sie Beiträge zur normalen Struktur, den Strukturwandlungen oder deren funktionellen Bedeutungen enthalten. Zum Publikationsgebiet des JOURNAL FÜR HIRNFORSCHUNG gehören auch Arbeiten, die sich mit der Zuordnung experimenteller Reiz- und Ausfallerscheinungen bzw. klinischen Symptomen zu bestimmten Strukturen des Gehirns („Lokalisationslehre") befassen. Als spezielles Publikationsgebiet ist die vergleichende Neurobiologie vorgesehen.

The JOURNAL FÜR HIRNFORSCHUNG will publish studies on normal morphology (anatomy, histology, cytology, electron microscopy, histochemistry), on the development of the nervous system, as well experimental anatomical studies. Papers of multidisciplinary character will also be included so far as they contain morphological results which were obtained using neuromorphological and neurophysiological or neuropharmacological and neurochemical methods and provide further information on the interaction between structure and function. Neuropathological studies will only be published of they contribute to the knowledge of normal structures structurals changes or their functional significance. Papers dealing with the cerebral localization of experimental excitation and deficit phenomena or clinical symptoms (localization theory) will also be published by the JOURNAL FÜR HIRNFORSCHUNG. A special part of the publication is reserved for comparative neurobiology.

Le JOURNAL FÜR HIRNFORSCHUNG publiera des études sur la morphologie normale (anatomie, histologie, cytologie, microscopie électronique, histochimie), sur le développement du système nerveux ainsi que des études anatomiques expérimentales. On acceptera aussi des travaux du caractère de la coopération entre des domaines différents à condition qu'ils contiennent des résultats morphologiques obtenus par les méthodes de la neuromorphologie et de la neurophysiologie ou de la neuropharmacologie et de la neurochimie. Les travaux doivent contenir des acquisitions nouvelles sur l'action réciproque entre la structure et la fonction. Des études neuropathologiques seront seulement acceptées quand elles contribuent à la conaissance des structures normales, des changements structurels ou de leur signification fonctionelle. Des études sur la localisation cérebrale de phénoménes expérimentaux ou cliniques d'excitation ou de déficit (doctrine des localisations) seront également publiées par le JOURNAL FÜR HIRNFORSCHUNG. Une partie spéciale sera réservée à la neurobiologie comparée.

Bezugsmöglichkeiten des „Journal für Hirnforschung".

Bestellungen sind zu richten

— in der DDR an eine Buchhandlung oder an den Akademie-Verlag,
DDR - 108 Berlin, Leipziger Straße 3–4

— im sozialistischen Ausland an eine Buchhandlung für fremdsprachige Literatur oder an den zuständigen Postzeitungsvertrieb

— in der BRD und Westberlin an eine Buchhandlung oder an die Auslieferungsstelle
KUNST UND WISSEN, Erich Bieber,
7 Stuttgart 1, Wilhelmstraße 4–6

— in Österreich an den Globus-Buchvertrieb,
1201 Wien, Höchstädtplatz 3

— im übrigen Ausland an den Internationalen Buch- und Zeitschriftenhandel; den Buchexport, Volkseigener Außenhandelsbetrieb der Deutschen Demokratischen Republik, DDR - 701 Leipzig, Postfach 160, oder an den Akademie-Verlag, DDR - 108 Berlin, Leipziger Straße 3–4

Journal für Hirnforschung

Herausgeber: Im Auftrag des Akademie-Verlages von einem internationalen Wissenschaftlerkollektiv herausgegeben.
Verlag: Akademie-Verlag, DDR - 108 Berlin, Leipziger Straße 3–4; Fernruf: 2236 221 oder 2236 229; Telex-Nr.: 114420; Bank: Staatsbank der DDR, Berlin, Kto.-Nr.: 6836-26-20712.
Schriftleitung: Prof. Dr. sc. med. W. Kirsche, Berlin.
Veröffentlicht unter der Lizenznummer 1326 des Presseamtes beim Vorsitzenden des Ministerrates der Deutschen Demokratischen Republik.
Gesamtherstellung: VEB Druckhaus „Maxim Gorki", DDR - 74 Altenburg.
Erscheinungsweise: Das Journal für Hirnforschung erscheint jährlich in einem Band mit 6 Heften. Bezugspreis eines Bandes 180,— M zuzüglich Versandspesen (Preis für die DDR 150,— M); Preis je Heft 30,— M (Preis für die DDR 25,— M).
Bestellnummer dieses Heftes 1018/19/5.

AN (EDV) 60315

ISSN 0021-8359

J. Hirnforsch. 19 (1978) 385–393

Journal für Hirnforschung

Internationales Journal für Neurobiologie

Heft 5 · 1978 · Band 19

Division of Neuropathology. University Medical Center Lausanne, Switzerland
(Head: Th. Rabinowicz, Assoc. Prof.)

Postnatal development of the mouse cerebral neocortex
IV. Evolution of the total cortical volume, of the population of neurons and glial cells*

By

Didier Heumann, Genevieve Leuba and Theodore Rabinowicz

With 3 figures and 2 tables

(Received August 24, 1977)

Kurztitel: Postnatal development of the neocortex

Summary: The total neocortical volume delimited between precise anatomical features was measured. Serial frontal sections cresyl-violet stained were measured with planimetric methods and the total volume evaluated at 5, 10, 30, 60 and 180 days. The total volume was used to obtain the total number of neurons and glial cells at each age using our quantitative data (published in our previous papers). The evolution of the cellular densities were studied in the sensory and motor areas 10–4–3 and 2 (Leuba and coll. 1977) and in the visual and auditory areas 17–18a–18–41 and 20 (Heumann and coll. 1977) in the mouse cerebral neocortex.

1. The total neocortical volume increases rapidly between 5 and 10 days, less between 10 and 30 days, slightly decreases at 60 days and reincreases at 180 days.
2. Between 5 and 30 days, the loss of cortical neurons is of about 30% by neuronal death.
3. The total number of glial cells doubles between 5 and 180 days. The increase is progressive through the development.

Résumé: Le volume néocortical, choisi entre des limites anatomiques précises, a été mesuré par planimétrie sur des coupes en série, fromtales, colorées au violet de crésyl, à 5, 10, 30, 60 et 180 jours. Il a servi à l'estimation du nombre total de neurones et de cellules gliales à ces âges à partir des données quantitatives sur l'évolution des densités cellulaires dans les aires sensorielles et motrices 10–4–3–2, Leuba et al. (1977) et dans les aires visuelles et auditives 17–18–18a–41 et 20, Heumann et al. (1977) du cortex cérébral de la souris.

1. Le volume néocortical augmente considérablement entre 5 jours et 10 jours, moins entre 10 jours et 30 jours, puis baisse légèrement à 60 jours et augmente à nouveau à 180 jours.
2. Il y a entre 5 jours et 30 jours une perte de neurones d'environ 30% dans l'écorce cérébrale, représentant une mort neuronale.
3. Le nombre de cellules gliales double dans le cortex cérébral entre 5 jours et 180 jours, augmentant sans cesse au cours de cette periode.

* This work was made possible trough grants from the Swiss National Science Foundation No. 3.641–0.71 and No. 3.434–0.74

Introduction

The total number of neurons and glial cells in the central nervous system is very likely the result of a balance between cell proliferation and cell death during neurogenesis, Glücksmann (1951), Källen (1955), Cowan (1973).

How does the total number of neurons evolve in the course of the cortical development? What are the relationships between the decreasing neuronal density and the increasing cortical volume? What is the evolution of the glial proliferation in the cortex if the glial density remains more or less unchanged till adult age? These are some of the questions we would like to answer using both our previous data, Leuba and coll. (1977), Heumann and coll. (1977) and those of the present research.

In two previous papers we provided quantitative data on the mouse cerebral neocortex during its development, Leuba and coll. (1977), Heumann and coll. (1977), and in a third paper we compared between themselves the evolution of neuronal densities and cortical thicknesses in some areas, Leuba and coll. (1978). As a matter of fact, the variations of neuronal densities during the development provides some clues on the maturation of the areas and of the cortical layers, but do not yield any information on the evolution of the total number of cells (neurons and glial cells) in the maturing cerebral cortex.

Thus we need to know the total number of cells in the cerebral cortex at different ages. This problem is much easier to investigate in clearly delimited structures and is now solved in the cochlear nucleus, Mlonyeni (1967), Königsmark and Murphy (1972), in the oculo-motor nucleus, Zilles and Wingert (1973), in the isthmo-optic nucleus, Clarke and coll. (1976), in the trochlear nucleus, Sohal (1976), in the dorsal geniculate nucleus, Heumann (not yet published data).

To our knowledge, it has never been done in the

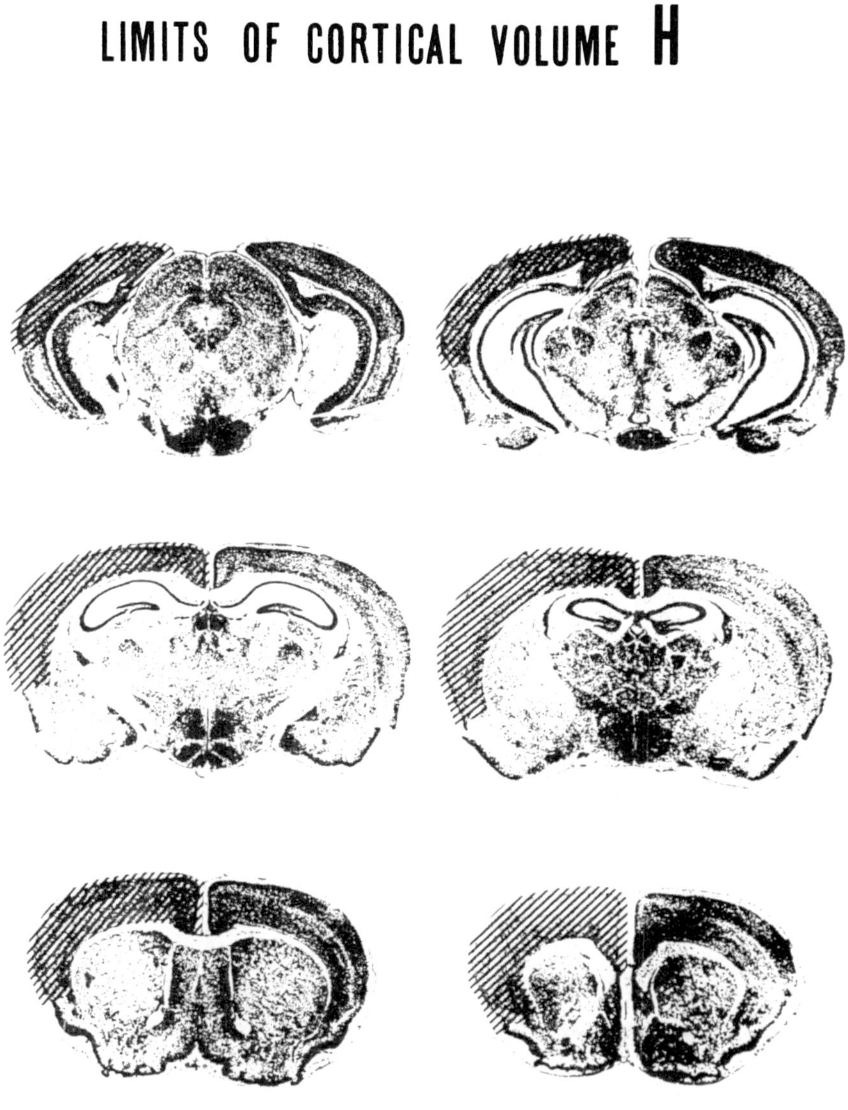

Fig. 1. Limits of cortical volume H. Upper left: first measured section in the caudorostral direction at the level of the appearance of the stratum granulosum of Ammon's Horn.
Lower right: last measured section in the caudorostral direction at the level of the end of the callosal interhemispheric fibers.
Inbetween, lateral limits at four different levels: laterally, rhinal fissure and medially, first subiculum and then interhemispheric fissure.

cerebral cortex. The problem is in fact very difficult because of the poor delimitation of the areas and of the layers. Nevertheless, we tried to avoid these difficulties by measuring a cortical volume between well recognizable anatomical limits arbitrarily chosen (D. HEUMANN) and by estimating the total number of neurons and glial cells in this volume.

Material and methods

I. Material

The same material is used as mentioned in our previous papers, LEUBA and coll. (1977), HEUMANN and coll. (1977).

II. Methods

1. Quantitative Cytoarchitectonics

The neuronal and glial densities have been obtained according to methods described and discussed in 2 previous papers and from the following areas 10—4—3—2—17—18—18a—41—20, KRIEG (1946), and at the following ages 5—10—30—60—180 days.

2. Cortical thickness

The whole brains of male Swiss Albinos mice are embedded in paraffin and cut serially and perpendicularly to the antero-posterior axis of the brain. The sections are 25 micra thick and stained with Cresyl-Violet using standardized methods LEUBA and coll. (1977). Ten measurements are made for each area in each of 10 animals.

3. Estimation of the cortical volume between arbitrarily chosen anatomical limits (volume H).

The measure or the cortical volume is made according to the method of KÖNIGSMARK (1970). Every fifth section is projected at a 50 X magnification. The cortical surface is measured with a planimeter (type Amsler 800/14807); the value obtained in micra^2 is multiplied by the thickness of the section (25 μ) and the number of sections and divided by the square of the magnification factor. These measures are performed on the basis of the following well recognizable anatomical criteria between arbitrarily chosen limits and called here volume H. These criteria are, laterally: rhinal fissure; medially: subiculum; caudally: appearance of the stratum granulosum of Ammon's horn; rostrally: end of the interhemispheric callosal fibers (fig. 1). Anteriorly and posteriorly to these limits the neocortex presents a less well organized structure not suitable for quantifications; laterally we find the paleocortex and medially the subiculum and the corpus callosum.

The volume H contains the above mentioned areas and represents most of the mouse's neocortex. The volume H was measured on the left hemisphere.

4. Estimation of the number of cells in a cortical parallelepiped

The cell densities have been obtained in each area and each layer for our maturational study. To achieve a realistic estimation of the total number of cells in the volume H, we use the number of cells obtained in each studied area. As it is impossible to obtain the total number of cells in a given area and as our data were obtained for each layer, we decided to consider a theoretical parallelepiped going vertically through all the layers as being representative of the whole area (see Fig 2). In order to obtain the best possible estimation of this parallelepiped using our data, the following operations have been done for each of the 10 brains studied at each age (to randomize our results 2).

a) refer the number of neurons and glial cells per unit of volume (0,00025 mm^3) in each layer and corrected by ABERCROMBIE (1946), to the total depth of the considered layer (fig 2).

$$N = A_{\mathrm{I}\cdots\mathrm{VI}} \times \frac{T_{\mathrm{I}\cdots\mathrm{VI}}}{78 \text{ micra}}$$

N = total number of cells in each layer on a 130 micra wide graticule after correction of ABERCROMBIE (1946).

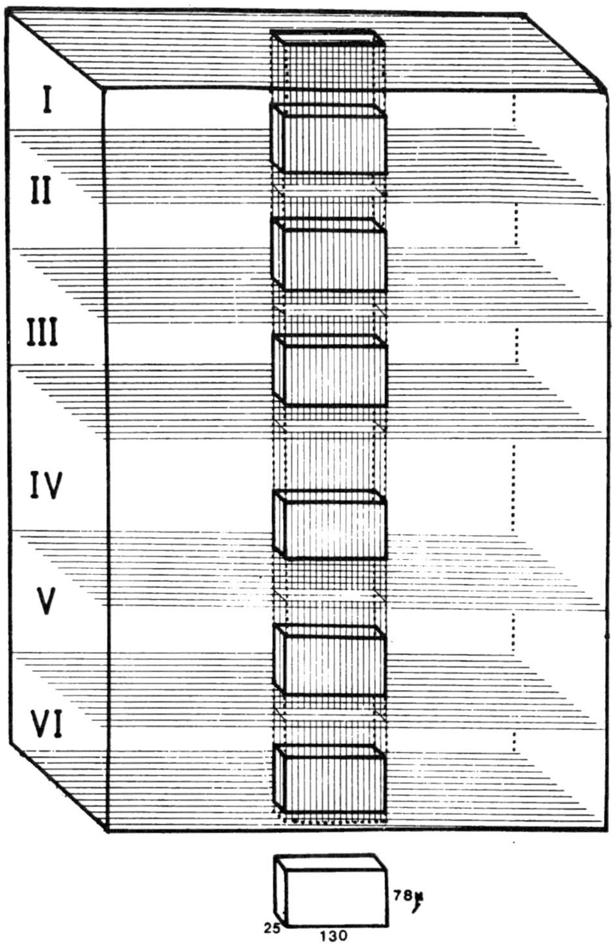

Fig. 2. Unitary parallelepiped: this volume is realized by considering that the cell densities found in the counting grids (volume constituted by the thickness of the section (25 micra), the length (130 micra) and the width (78 micra) of the graticule) in each layer represent an average unit for establishing the theoretical volume of the unitary parallelepiped. Thus is parallelepiped goes through the whole depth of the cortex and produces a deep and narrow volume whose theoretical basis is a surface of 25 μ × 130 μ (one side of the graticule (130 μ) × thickness of the section (25 μ), with a height constituted by the depth of the cortex.

$A_{I\ldots VI}$ = number of cells per unit of volume (0.000 25 mm^3) after ABERCROMBIE'S (1946) correction, in layers I to VI.

$T_{I\ldots VI}$ = thickness of each layer.

78 micra: height of the graticule. This value correspond to the vertical sampling dimension of our graticule (Fig. 2).

b) Add the values from formula a) for all the six layers to obtain the theoretical number of cells in the total cortical depth in each area.

$B = N_{I\ldots VI}$ = number of cells in the whole cortical depth on a width of 130 micra and for a given area (the width of the graticule corresponds to 130 micra).

c) Refer the value obtained under b) to a theoretical thickness of 1 000 micra that we have chosen in order to obtain a cortical unitary parallelepiped identical in all areas because of the variable depths of the different areas constituting our volume H (fig. 2).

$B_{cor} = \frac{B. \times 1\,000 \text{ micra}}{\text{depth of the area}}$ = total number of cells in a unitary parallelepiped of 130 micra $\times$ 1 000 micra $\times$ 25 micra = 0.003 25 mm^3 for each area.

5. Estimation of the total number of neurons and glial cells in volume H.

We calculate the mean of all the numbers obtained in the parallelepipeds (B cor) for all the 9 areas of each of the 10 animals. This will represent the average number of cells for each unitary parallelepiped in each brain (P un. = 0,003 25 mm^3). We obtain the estimation of the number of neurons or glial cells by refering for each brain the number of cells obtained in the unitary parallelepiped to its total cortical volume H.

Total number of cells: number of cells in P un. $\times \frac{\text{volume } H}{0{,}003\,25 \text{ mm}^3}$

Finally we average the results for the 10 brains at each age.

Results

I. Cortical Volume H (table 2)

The cortical volume H increases in a considerable way between 5 and 10 days and slightly less between 10 and 30 days. The differences are highly significant in the Student's test ($p < 0{,}001$). From 30 days on, some fluctuations appear particularly a slight decrease at 60 days compared to 30 days followed by a slight increase at 180 days compared to 60 days. The cortical volume H at 30 days is the highest of all, higher even than at 180 days. The phenomenon is similar to what has been described for the cortical thickness, LEUBA and coll. (1978).

II. Relations between cortical volume, cortical thickness and neuronal density

In our 3 previous papers, LEUBA and coll. (1977), HEUMANN and coll. (1977), LEUBA and coll. (1978) several cortical features appeared in the course of the postnatal maturation:

Between 5 and 10 days:

1) an important decrease of the neuronal density always more important for layers II, III and IV than for layers I, V and VI. .

2) a sharp increase of the depth for all the layers.

3) a great increase of the cortical volume H (table 2).

Between 10 and 30 days:

1) a slighter decrease of the neuronal density always more pronounced in layers II, III and IV than in layers I, V and VI.

2) an increase in depth of all the layers.

3) an increase of the cortical volume H (table 2).

In fact until 30 days the neuronal density decreases in all the layers with different speeds depending on the layers. Inversely both cortical depth and cortical volume H increase. These events represent well the cortical development. The decrease of the number of neurons by unit of volume is correlated to the increase of the neuropile as well as to the glial and endothelial proliferation. The question now is whether the correlation is exact or if any other phenomena may interfer, particularly the cellular death.

From 30 days on: the neuronal density is more or less stabilized in the nine studied areas. However, generally the density is slightly higher at 60 and 180 days than at 30 days. As a matter of fact a higher neuronal density is correlated with the decrease in cortical thickness and cortical volume at 60 and 180 days compared to 30 days.

Cellular density on one side and cortical depth and cortical volume on the other side are thus bound at each stage of the development.

III. Estimation of the total number of cells

We have seen that during the development when the cortical volume increases, the neuronal density decreases. But we did not know if one phenomenon counterbalanced exactly the other, nor did we know if the total number of cells changed. That is why we tried to estimate the total number of neurons and glial cells at each age (5—10—30—60—180 days) in the cortical volume H.

1. Estimation of cells in a unitary cortical parallelepiped (table 1)

Let us recall that this measure is necessary to the estimation of the total amount of neurons and glial cells in the volume H. As shown in table 1 the variations of the neuronal density in the parallelepiped of

Table 1: Neuronal densities in the theoretical parallelepiped of cortical volume in areas 10, 4, 3, 2, 17, 18, 18a, 41 and 20 (KRIEG 1946), at 5, 10, 30, 60 and 180 days.
Mean values; extremes in parenthesis.
Neuronal density in 0,003 25 mm³ of cortex

	5 days	10 days	30 days	60 days	180 days
area 20	1928 (1424–2380)	759 (676– 878)	478 (422–625)	526 (440–680)	534 (428–616)
area 41	2213 (1420–2959)	765 (665– 864)	520 (445–660)	562 (452–661)	547 (470–627)
area 18a	2244 (1463–2748)	830 (679– 942)	522 (444–654)	561 (449–639)	544 (474–596)
area 18	1833 (1478–2781)	748 (612– 898)	536 (509–590)	589 (415–657)	549 (489–622)
area 2	2238 (1972–2654)	883 (768– 996)	504 (483–523)	529 (495–586)	519 (475–571)
area 17	2661 (1607–3297)	1006 (853–1364)	590 (535–702)	617 (554–680)	630 (533–718)
area 3	2373 (1931–2887)	872 (769–1062)	513 (473–556)	572 (546–614)	557 (478–626)
area 4	2090 (1734–2486)	801 (715– 955)	476 (447–509)	517 (488–563)	509 (445–569)
area 10	1841 (1165–2366)	715 (619– 898)	383 (350–458)	404 (360–445)	405 (317–475)

Table 2. Cortical volumes H, total numbers of neurons and glial cells in volume H, at 5, 10, 30, 60 and 180 days.
Mean values, extremes in parenthesis.

Cortical Volumes H in mm³

5 Days:	7,276 (5,469– 9,012)
10 Days:	14,610 (11,126–18,048)
30 Days:	21,659 (17,770–23,425)
60 Days:	19,248 (16,623–21,713)
180 Days:	20,532 (16,022–22,825)

Total Number of Neurons

5 Days:	4.798.462 (4.373.538–5.351.385)
10 Days:	3.634.769 (3.160.923–4.129.538)
30 Days:	3.352.307 (3.154.769–3.513.230)
60 Days:	3.208.000 (2.874.461–3.519.385)
180 Days:	3.441.231 (2.733.846–3.888.000)

Total Number of Glial Cells

5 Days:	426.984 (369.077–495.231)
10 Days:	590.707 (402.923–726.369)
30 Days:	789.723 (663.231–879.477)
60 Days:	755.785 (609.446–883.938)
180 Days:	817.815 (627.569–934.738)

0,003 25 mm³ show in all the areas the same trends: high at 5 days, lower 10 days and rather stable from 30 days on. Still some differences between areas exist: ganular areas (17, 2, 3) show at all ages a higher mean density due to the denser layer IV in these areas. Areas 20 and 18 which have no evident layer IV, have a lower mean density, specially at 5 and 10 days. Area 10 has always the lowest cell density. Between 30 and 180 days, the mean densities in all our areas show a tendency to level.

1. Total number of neurons in the volume H (table 2, fig. 3)

The estimation of the number of neurons in our cortical parallelepiped at each age does not indicate any great differences between areas. We are thus allowed to consider the mean of all areas as enough representative of the whole neocortex. Knowing the total volume H and the mean neuronal density in the unitary volume of 0,003 25 mm³, we can determine the total number of neurons in volume H.

Remarkably enough, the total number of neurons, far from being stable, decreases between 5 and 10 days of about 24%; the decrease is less between 10 and 30 days, about 6%.

After 30 days there are only slight fluctuations of less than 5%.

3. Total number of glial cells in volume H (table 2; fig. 3)

The total number of glial cells is obtained in the same way as for the neurons.

We found a constant increase of the total glial population during the postnatal development, important till 30 days, decreasing slightly at 60 days and increasing at 180 days even more than at 30 days.

The glial proliferation is thus a rather important phenomenon as it doubles between 5 and 180 days.

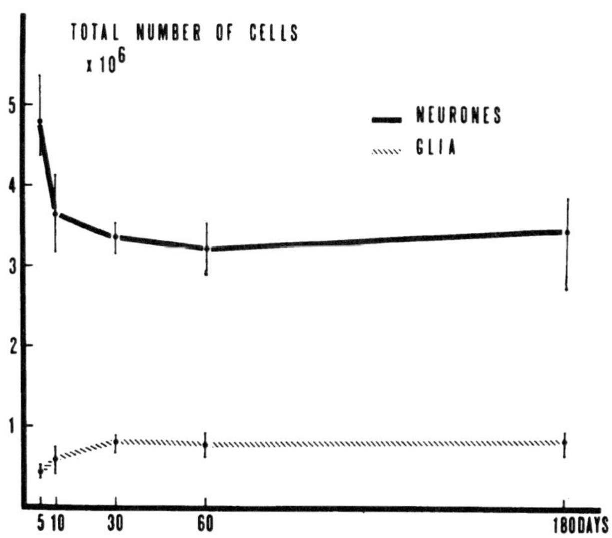

Fig. 3. Evolution of the total number of neurons and glial cells between 5 and 180 days in the cortical volume H. Each point represents the mean value taken from 10 different animals. Extremes indicated.

Discussion

I. Neuronal death

In our estimation of the total number of neurons of volume H, we found a decrease of about 30% between 5 and 30 days. This loss of 30% of neurons during the postnatal maturation represents likely a neuronal death, although we have never observed any clear sign of degeneration or necrosis, picnotic nuclei or ghost cells.

The concept of cellular death and more specifically neuronal death, is relatively old in embryology, HAMBURGER and LEVI-MONTALCINI (1949). It has been confirmed by GLÜCKSMANN (1951) and KÄLLEN (1955).

More recently COWAN (1973) has made a review of this notion during the growth of the CNS. Most of the papers are dealing with chicken and Xenopus laevis material and present evidence for neuronal death during the development. Recently CLARKE and coll. (1976) have shown a neuronal loss going up to 60% in the isthmo-optic nucleus and SOHAL (1976) has shown the same phenomenon in the trochlear nucleus. In the motor horns of Xenopus laevis, PRESTIGE (1967) has found that the final neuronal population was 1/4 of the initial one.

In mouse also some authors are in favour of a neuronal death during the maturation. ROMANES (1946) has shown a loss of neurons of about 30% in the ventral horn of the spinal cord between 2 and 12 postnatal days. MLONYENI (1967) has found a loss of neurons in the cochlear nucleus ($\sim$ 30%) between 9 and 16 postnatal days. ZILLES and WINGERT (1973) evidenced a loss of 30% of neurons in the oculomotor nucleus between birth and the 20th postnatal day. SCHLESSINGER and coll. (1975) in an autoradiographic study of the gyrus dentatus in rat, have suggested a cellular death of about 40%. Moreover in the lateral geniculate nucleus of mouse (D. HEUMANN, no yet published data) there is again a neuronal loss of about 30% between 5 and 30 days postnatal.

To our knowledge these phenomena have not yet been quantitatively demonstrated in the isocortical areas in mice or other mammals.

It is easier to ascertain results in the lateral geniculate nucleus or in the oculo-motor nucleus because these nuclei are well limited and represent homogenous structures. Consequently their total number of cells can be more precisely evidenced. The cerebral cortex as a whole is not an easy limitable structure. Our volume H is only an estimation of the neocortical one (see techniques) and we must discuss here at least 2 points:

1. Is this volume H representative of the real neocortical volume and what is the precision of our methods for its estimation?

As already shown (see techniques) our volume H represents most of the neocortical areas of the mouse. The anterior and posterior parts of the neocortex have been eliminated because of their poor limits with the paleocortex. Another fact is the possibility that the chosen anatomical limits do not remain exactly at the same place during the development along the antero-posterior axis of the brain, introducing thus a small percentage of error on the contrary. The lateral and the medial limits are easily recognizable and safe.

Moreover KÖNIGSMARK (1970) showed that if one section only out of five or more is used, an error of about 5% is introduced into the method. As this error is present in all the brains and at all ages, it can been considered as a constant in our research which is mostly a comparative one. On the other hand the error introduced by the planimetric method is considered as really negligible (less than 0,5%). We thus consider our volume H as being well representative of the real neocortical volume.

2. The second question is: Are the samples of our areas enough representative of the mean density of the cerebral cortex?

As it is almost impossible to count every cell in the total cerebral cortex even in a mouse, the best method would have been to count solid columns of cells from the subpial surface down to the white matter in the different areas of the cortex. A mean of all these countings would also give an average cell density of the whole cerebral cortex. Instead of this we used our previous data and we averaged these countings made on separate layers in the same area so that we obtained a theoretical column of cells. We used nine areas representative in our opinion of the whole neocortex. As already shown (table 1) the differences of cell densities in each theoretical parallelepiped of 0,00325 mm^3 are not so different between one area and another. An average value for the whole neocortex is thus deducible.

As for the cortical volume the error in such estimation is the same in all the brains and through all the different stages of the development. Thus it does not much interfere with the comparability between different ages and can not explain the 30% loss of cells appearing between 5 and 30 days. As a consequence we assume that this phenomenon represents surely neuronal death.

Moreover our results are well in accordance with research done on other neural structures.

COWAN (1973) has suggested an overproduction

of neurons during maturation, this mecanism being a sort of regulation to eliminate neurons without functional connections. Indeed the maximum of cellular death happens often at the moment when most of the connections with the periphery are established. The brain of the newborn mouse is very poor in connections and the dendrites begin to grow after birth, HADDARA (1956), EAYRS and GOODGEAD (1959), MELLER and coll. (1968). Our GOLGI study (RABINOWICZ and coll. (1977)) shows the same phenomenon. Thus it is very likely that a competition occurs during the establishment of connections. DAWKINS (1971) mentioned even that a selective neuronal death could be the support of a mnemonic mecanism.

II. Glial proliferation

The density of glial cells varies very little in the course of the postnatal development in our nine areas, LEUBA and coll. (1977), HEUMANN and coll. (1977). It remains rather steady at each age in the same layer. Deep layers have always slightly more glial cells than superficial ones, layer I excepted. The glial density, slightly higher at 5 days, is likely due to the extremely high nuclear density of the cerebral cortex at this age, which makes a confusion always possible with nuclei of small neuroblasts. Thus we can not assure that there are clearly recognizable glial cells, but there are anyway "glial-like" cells at 5 days. The first glial cells assigned to the cerebral cortex are born around birth, BERRY and ROGERS (1965), HICKS and D'AMATO (1968), BRÜCKNER and coll. (1976). Thus a certain number of them could already be in the cortical layers at 5 days. In the more mature layers I, V and VI the glial cells can be distinguished from the neurons with enough precision. In layer II which is very dense we have not been able to count them and in layers III and IV the discrimination is not safe enough.

From 10 days on, there are no more problems of identification and till 180 days the density is more or less a steady one in each layer.

If we consider now the total number of glial cells in the volume H (table 2) a real glial proliferation occurs during the cortical maturation. It goes from about 425'000 at 5 days to 590'000 at 10 days and to 820'000 at 180 days, which represents an important increase.

This increase is specially important till 30 days and follows the increase of the cortical volume. At 60 days the volume is slightly smaller and the number of glial cells also. At 180 days the volume is again slightly higher and so is the number of glial cells. However the cortical volume H is at its maximum at 30 days and the number of glial cells at 180 days. Between 30 and 180 days the increase in glial cells is of 12—13%. These results allow us to draw two conclusions:

1) There is a steady glial proliferation in the cerebral cortex of the mouse at least till 180 days (our study does not go further for the moment).

2) There seems to be a fixed relationship between the total number of glial cells and the cortical volume they occupy.

In the white matter we could not approach the question of the total number of glial cells because of its very difficult delimitation. STURROCK (1974) showed in the well delimited anterior horn of the anterior cerebral commissure (in mouse) that the glial density is more or less stable during the maturation. As he could measure the total volume of the structure it appeared that the total number of glial cells progresses from 225 at 16 embryonic days to 11'000 at 30 postnatal days with a maximum increase between 8 and 14 days. Thus the glial proliferation was indeed parallel to the increase of volume. This corresponds well to what we have shown in the cerebral cortex. Consequently it is very likely that the same phenomenon appears also in the white matter of the hemispheres.

The glial proliferation during the postnatal cortical development of mouse and even at adult age is a well known mechanism, BRIZZEE and al. (1964; 1968), LING and LEBLOND (1973). Moreover it has also been evidenced by numerous autoradiographic studies, among others DALTON and coll. (1968), KRAUS-RUPPERT and coll. (1973), KORR and coll. (1973), MARES and LODIN (1974) etc.

Our results confirm with a quantitative neuroanatomical method that a considerable glial proliferation happens during the maturation and goes on till adult age (180 days) in the mouse's neocortex.

III. Relationships between cortical thickness, cortical volume and cellular density

The neuronal density on one side, the thickness of the layers or the cortical volume and the glial proliferation on the other side, vary in a reversed way. The increase of the cortical depth and volume and the correlated decrease of neuronal density represent a possibility to appreciate the cortical maturation. Till 30 days the cortical thickness and the cortical volume increase, together with the neuropile RABINOWICZ and coll. (1977), while the neuronal density decreases progressively.

But the two phenomena are not exactly proportional because of the neuronal death till 30 days. The neuronal density can be considered as rather stabilized in spite of some statistically significant differences between 30 and 60 days or 60 and 180 days: the number of neurons by unit of volume is slightly higher at 60 and 180 days than at 30 days. Remarkably enough at 60 days the cortical depth and the cortical volume are slightly smaller.

Diamond and coll. (1975) have shown in the rat that the cortical thickness increases till about 26 days, reaching a maximum somewhere between 26 and 40 days and then decreases till 108 days being stabilized from that moment. Our data show equally a maximum of cortical thickness (and cortical volume) at 30 days. So the cerebral cortex of the mouse may have a tendency to decrease slightly in thickness after having reached a maximum in the course of the maturation.

Aknowledgements

Mrs. M. Meier, Mrs. R. Lachat and Miss Z. Trbovic prepared the microscopical slides; Mr. B. Maurer did the photographic work; Miss L. Deprez typed the manuscript.

Bibliography

Abercrombie, M.: Estimation of nuclear population from microtome sections. Anat. Rec. **94**, 239—247 (1946).

Berry, M. and A. W. Rogers: The migration of neuroblasts in the developing brain cortex. J. Anat. (Lond.) **99**, 691—709 (1965).

Brizzee, K. R., J. Vogt and X. Kharetchko: Postnatal changes in glial neuron index with a comparison of methods of cell enumeration in the white rat. In: Progress in Brain Research **4**, 136—149 (1964), Purpura and Schadé (eds.) Elsevier, Amsterdam.

Brizzee, K., N. Sherwood and P. S. Timiras: A comparison of cell populations at various depth levels in cerebral cortex of young adult and aged Long Evans Rats. J. Gerontol. **23**, 289—297 (1968).

Brückner, G., V. Mares and D. Biesold: Neurogenesis in the visual system of the rat. An autoradiographic investigation. J. Comp. Neurol. **166**, 245—256 (1976).

Clarke, P. G. H., L. A. Rogers and W. M. Cowan: The time of origin and the pattern of survival of neurons in the isthmo-optic nucleus of the chick. J. Comp. Neurol. **167**, 125—142 (1976).

Cowan, W. M.: Neuronal death as a regulative mechanism in the control of cell number in the nervous system. In: Development and Aging in the Nervous System, 19—41 (1973). M. Rockstein (ed.) Academic Press, New York.

Dalton, M. M., O. R. Hommes and C. P. Leblond: Correlation of glial proliferation with age in the mouse brain. J. Comp. Neurol. **134**, 397—400 (1968).

Dawkins, R.: Selective neuronal death as a possible memory mechanism. Nature **229**, 118—119 (1971).

Diamond, M. C., R. E. Johnson and C. A. Ingham: Morphological changes in the young, adult and aging rat cerebral cortex, hippocampus and diencephalon. Behav. Biol. **14**, 163—174 (1975).

Eayrs, J. T. and B. Goodhead: Postnatal development of the cerebral cortex in the rat. J. Anat. (Lond.) **93**, 385—402 (1959).

Glücksmann, A.: Cell death in normal vertebrate ontogeny. Biol. Rev. **25/26**, 59—86 (1951).

Haddara, M.: A quantitative study of the postnatal changes in the packing density of the neurons in the visual cortex of the mouse. J. Anat. (Lond.) **90**, 494—501 (1956).

Hamburger, V. and R. Levi-Montalcini: Proliferation, differentiation and degeneration in the spinal ganglia of the chick embryo under normal and experimental conditions. J. Exp. Zool. **111**, 457—500 (1949).

Heumann, D., G. Leuba and Th. Rabinowicz: Postnatal development of the mouse cerebral neocortex. II. Quantitative cytoarchitectonics of visual and auditory areas. J. Hirnforsch. **18**, 483—500 (1977).

Hicks, S. P. and C. J. d'Amato: Cell migrations to the isocortex in the rat. Anat. Rec. **160**, 619—634 (1968).

Källen, B.: Cell degeneration during normal ontogenesis of the rabbit brain. J. Anat. (Lond.) **89**, 153—160 (1955).

Königsmark, B. W.: Methods for the counting of neurons. In Contemporary Research Methods in Neuroanatomy, 315—340 (1970), Nauta and Ebesson (eds), Springer-Verlag.

Königsmark, B. W. and E. A. Murphy: Volume of the ventral cochlear nucleus in man: its relationship to neuronal population and age. J. Neuropath. Exp. Neurol. **31**, 304—316 (1972).

Korr, H., B. Schultze and W. Maurer: Autoradiographic investigation of glial proliferation in the brain of adult mice I. The DNA synthesis phase of neuroglia and endothelial cells. J. Comp. Neurol. **150**, 169—176 (1973).

Kraus-Ruppert, R., J. Laissue, Burki, H., and N. Odartchenko: Proliferation and turnover of glial cells in the forebrain of young adult mice as studied by repeated injections of *3H*-thymidine over a prolonged period of time. J. Comp. Neurol. **148**, 211—216 (1973).

Krieg, J. S.: a) Connections of the cerebral cortex. I. The albino rat. A. Topography of the cortical areas. J. Comp. Neurol. **84**, 221—276 (1946).
b) Connections of the cerebral cortex. I. The Albino rat. B. Structures of the cortical areas. J. Comp. Neurol. **84**, 277—322 (1946).

Leuba, G., D. Heumann and T. Rabinowicz: a) Postnatal development of the mouse cerebral neocortex. I. Quantitative cytoarchitectonics of some motor and sensory areas. J. Hirnforsch. **18**, 461—481 (1977).
b) Postnatal development of the mouse cerebral neocortex. III. Some dynamical aspects. J. Hirnforsch. **19**, 301—312 (1978).

Ling, E. A. and C. P. Leblond: Investigation of glial cells in semithin sections. II. Variation with age in the number of the various glial cell types in rat cortex and corpus callosum. J. Comp. Neurol. **149**, 73—82 (1973).

Mares, V. L. and Z. Lodin: An autoradiographic study of DNA synthesis in adolescent and adult mouse forebrain. Brain Res. **76**, 557—561 (1974).

Meller, K., W. Breipohl and P. Glees: Synaptic organization of the molecular and the outer granular layer in the motor cortex in the white mouse during postnatal development. A Golgi and electron microscopical study. Z. Zellforsch. **92**, 217—231 (1968).

MLONYENI, M.: The late stages of the development of the primary cochlear nuclei in mice. Brain Res. **4**, 334–344 (1967).

PRESTIGE, M. C.: The control of cell number in the lumbar ventral horns during the development of Xenopus Laevis tadpoles. J. Embryol. Exp. Morph. **18**, 359–387 (1967).

RABINOWICZ, Th., G. LEUBA and D. HEUMANN: Morphological maturation of the Crain: A qualitative study. In: Brain, fetal and infant 28–53 (1977) Berenberg (ed.), M. NIJHOFF, The Hague.

ROMANES, C. J.: Motor localization and the effects of nerve injury on the ventral horn cells of the spinal cord. J. Anat. (Lond.) **80**, 117–131 (1946).

SCHLESSINGER, A. R., W. M. COWAN and P. I. GOTTLIEB: An autoradiographic study of the time of origin and the pattern of granule cell migration in the dentate gyrus of the rat. J. Comp. Neurol. **159**, 149–176 (1975).

SOHAL, G. S.: An experimental study, of cell death in the developing Trochlear nucleus. Exp. Neur. **51**, 684–698 (1976).

STURROCK, R. R.: Histogenesis of the anterior limb of the anterior commissure of the mouse brain. I. A quantitative study of changes in the glial population with age. J. Anat. (Lond.) **117**, 17–25 (1974).

ZILLES, K. and F. WINGERT: Quantitative studies of the development of the fresh volumes and the number of neurons of the nuclei N. oculo-motori of white mice during ontogenesis. Brain Res. **56**, 63–75 (1973).

Adress of the authors:

Th. RABINOWICZ, M. D.
Assoc. Prof.
Division of Neuropathology
Centre Hospitalier Universitaire Vaudois
1011 Lausanne
Switzerland

Buchbesprechung

Advance in Neurosurgery, Vol. 4

Lumbar Disc
Adult Hydrozephalus
Editors: Wüllenweber, R.; Brock, M.; Hamer, J.; Klinger, M.; Spoerri, O.
154 figs., 67 tables. XXIII, 338 pages (8 pages in German) 1977.

Die während der 27. Jahrestagung der Deutschen Gesellschaft für Neurochirurgie gehaltenen Vorträge vermitteln einen Überblick über den derzeitigen Stand der Behandlung von lumbalen Bandscheibenschäden sowie über die Hydrozephalusbehandlung im Erwachsenenalter. Freie Themen runden den Band ab. Schemata, Abbildungen sowie angefügte Literaturangaben ermöglichen den Erwerb weiterer Informationen. Druck und Qualität der Abbildungen sind von gewohnter, guter Qualität.

Der Wert der diagnostischen, differentialdiagnostischen und operativen Aussagen ist in dem großen, teilweise computermäßig aufgearbeiteten Krankengut von Patienten mit lumbalen Bandscheibenschäden zu sehen. Beispielsweise ergab die CT-Aufschlüsselung der Frage, welche Rolle Anomalien bei der Diagnostik dieser Schäden spielen, daß Anomalien des Duralsackes, der Wurzeltaschen und/oder der epiduralen Gefäße in etwa 6,8% Ursachen der Symptomatik waren. Die Sammelstatistik von 3238 an 15 Neurochirurgischen Kliniken operierten Patienten mit Bandscheibenschäden zeigte, daß postoperativ in etwa 80% gute Ergebnisse zu erwarten sind. Anlaß zu Reoperationen sind überwiegend narbige Verklebungen. Nach Reoperationen ist in etwa 68% mit zufriedenstellenden Resultaten zu rechnen (Schlarb und Wenker, Oppel et al., Scheppelmann et al.). Das Operationstrauma wird bei Anwendung des Mikroskopes und der mikrochirurgischen Technik — Präzision des Eingriffes (Yasargil) — erheblich reduziert. Katamnestische Betrachtungen über psychische Symptome im Gesamtgeschehen eines Ischiassyndroms sowie gutachterliche Hinweise runden diesen Komplex ab (Spizing et al., Frowein und Terhaag). Im Komplex der Hydrozephalusbehandlung im Erwachsenenalter konzentrieren sich die Fragen auf die Behandlung des Normaldruckhydrozephalus. Für die Indikationsstellung zur Ventiloperation werden neben den üblichen klinischen und neuroradiologischen Maßnahmen langzeitige szintigraphische Untersuchungen sowie fortlaufende Registrierungen des Liquordruckes gefordert. Die CT-Kontrolle ermöglicht Aussagen zum Operationseffekt. Weder geben die experimentellen noch die klinisch-morphologischen Ausführungen — hinsichtlich der Liquorkinetik — sichere Aufschlüsse.

Innerhalb der „Freien Themen" nehmen die Schädel-Hirn- und Rückenmarkverletzungen einen breiten Raum ein. EEG-Überwachungen sowie computerisierte EEG-Analysen stellen in der Frühphase nach Schädel-Hirnverletzungen eine gute Grundlage für weitere therapeutische Entscheidungen, aber auch für prognostische Aussagen dar (Miltner und Soerensen, Krüger et al.). Nach Terhaag ist die meist als „günstig" bezeichnete Prognose dieser Verletzungen im Kindesalter zweifelhaft. Aktuell sind die Fragen der Dexamethasonbehandlung des Hirnödems. Hohe Dosen scheinen sich sowohl beim posttraumatischen, aber auch tumorbedingten Hirnödem zu bewähren (Faupel et al., Wiegand et al.). Der Therapieeffekt kann mittels CT und langzeitigen intrakraniellen Druckmessungen verfolgt bzw. bestätigt werden. Die Ableitung evozierter Potentiale nimmt im Rahmen der Beurteilung spinaler und supraspinaler Läsionen eine gefestigte Position ein. Interessante Aspekte ergeben sich mit der Elektrostimulation mittels implantierbarer Elektroden bei spastischen Zuständen. (Kraunik et al.)

Weitere experimentelle Arbeiten sind der mikrochirurgischen Rekonstruktion von Cauda-equina-Fasern und der Anastomosetechnik spinaler Gefäße (Kletter et al.) gewidmet.

Die von Mundinger und Hoefer angegebene Methode zur Lokalisation des Foramen Monroi sowie die Vorteile des CT-Einsatzes bei stereotaktischen Eingriffen sind erwähnenswert (Birg, Mundinger und Klar). Mit dem Nachweis „gemischter Areale" ließ sich anatomisch die Kommunikation zwischen arteriellen und venösen Gefäßsystem im Kleinhirnbereich sichern.

Der Sammelband ist für die Tätigkeit in der Sprechstunde ebenso empfehlenswert wie für die Klinik.

(Unger, Berlin)

J. Hirnforsch. 19 (1978) 395–403

From the Department of Physiology, University Medical School, Pécs, Hungary

Divergent responses to thyroid hormone treatment of the different secondary germinal layers in the postnatal rat brain

By László SERESS

With 11 figures

(Received)

Key words:
Thyroid, cerebellum, germinal layer, dentate gyrus, hypothalamus, cell formation.

Summary: The effect of daily triiodothyronine (T_3) treatment (first injection on the day of birth) was studied on the postnatal development of various parts of the rat brain.
It was found that the T_3 treatment acts differently upon the postnatal cell formation in the various secundary germinal layers during the first postnatal week. The T_3 treatment resulted in an increase of the cell multiplication in the external granular layer of the cerebellum but decreased the cell division or had no significant effect in the periventricular germinal layer and in the polymorph layer of the dentate gyrus.
From the 10th day the T_3 treatment resulted in a decrease of the cell division in all secondary germinal layers examined.
As a reason for this different effect it can be suggested that the triidothyronine acts differently upon the various neuronal and glial precursors or upon the germinal layers producing them.

Zusammenfassung: Es wurde der Effekt einer täglichen Trijodthyronin-Applikation (T3) (erste Injektion am Tage der Geburt) auf die postnatale Entwicklung verschiedener Teile des Gehirns untersucht.
In der ersten Lebenswoche wirkt die T3-Behandlung unterschiedlich auf die postnatale Zellbildung in den verschiedenen sekundären Matrixzonen.
T3-Behandlung führt zu einer erhöhten Zellbildung im Stratum granulosum externum des Cerebellum, aber zu einer verminderten Zellteilung bzw. zu einem nicht signifikanten Effekt im Stratum periventriculare und im Stratum multiforme der Fascia dentata.
Nach dem 10. Tag der T3-Behandlung resultiert eine verminderte Zellteilung in allen untersuchten sekundären Matrixschichten. Als Ursache dieser unterschiedlichen Effekte kann angenommen werden, daß Trijodthyronin unterschiedlich auf die Nerven- und Gliazellen im frühen Stadium ihrer Differenzierung wirkt oder auf die Matrixzonen, aus denen sie hervorgehen.

Introduction

It is generally believed that thyroid hormone inhibits the postnatal cell formation in the mammalian cerebrum and cerebellum (HAMBURGH, 1968; BALÁZS et al., 1971; NICHOLSON et al., 1972; KOVÁCS, 1973). However, a stimulating effect could also be demonstrated in the cerebellum by the administration of thyroid hormone in early period of life (GOURDON et al., 1973; KOVÁCS et al., 1975). This suggests that the response to thyroid hormone of various deeveloping structures of the brain may vary according to dose and/or the area studied.

The present work ist a comparative histological and autoradiographic study of the different secondary germinal layers of the brain of infant rats exposed to thyroid hormone.

Material and Methods

Litters of male albino (CFY) rats born on the same day were divided into experimental and control groups. Both groups contained animals of the same mother. Each group consisted of eight animals.

The animals were kept in 12 hours dark-light periods, and were given standard lab-chow and water ad libitum.

The animals of the experimental group were injected with daily 5 μg triiodothyronine (T_3) (solved in 0.05 ml isotonic saline) from the day of birth. Controls were injected with 0.05 ml isotonic saline solution. The last injection was given on the day of sacrifice.

One hour before their sacrifice animals received 10 μCi/g body weight 6-^{3}H-thymidine (specific activity: 18.4 Ci/mM) subcutaneously. The rats were killed by decapitation. The brains were fixed in 10% formaldehyde and embedded in paraffin. Serial sections of 5 μ thickness were cut in the frontal plane, covered with Kodak AR-10 stripping film or Ilford L4 Nuclear Emulsion. The slides were kept in light-proof boxes at 4 °C for six weeks when the autoradiograms were developed.

The slides were stained with gallocyanine or with the Nissl method. Evaluation was made at × 1200 magnification under oil immersion with the aid of a net reticule placed in one eyepiece of the binocular.

Nuclei containing more than ten grains were considered as labelled. The percentage figures for labelled cells referred to populations of 3–5,000 cells in each layer and in each animal.

The animals were sacrificed on postnatal days 2, 3, 5, 7, 10, 14, 18 and 21. Day 1 was the day of birth.

Each group consisted of three animals. Values were indicated on plot diagrams representing the mean of a group. The cell populations compared were fairly homogeneous in size thus the counting of cell profiles in sections yielded figures proportionate to packing density.

By the statistical evaluation the analysis of variance (and Student s t-test) was used, and ± means the standard error of mean. The external and internal granular layer and the white matter of cerebellum, the granular and polymorph layer of the dentate gyrus, the periventricular germinal layer and the hypothalamus were examined.

Fig. 2. Labelling indices in the EGL of the cerebellum. 1 hour labelling with 6-H^3-thymidine.
* ($p < 0.05$)
C: control
T_3: treated with triiodothyronine

Results

Cerebellum

As revealed by autoradiography labelled cells could be seen in all layers of the postnatal cerebellum (Fig. 1).

In the external granular layer (EGL) of the cerebellum 24% of the cells were labelled on the postnatal day 2 (Fig. 2). This value decreased on the third day and remained constant till the 14th day of life. The number of labelled cells decreased rapidly from the 14th postnatal day. On the 21st day, only 2% of cells were labelled.

The T_3 treatment resulted in a significant increase of labelling on the 5th day. There was no difference on the 7th and 10th day, but a significant decrease was found on the 14th and 18th day. The external granular layer could not be found on the 21st day in the animals treated with T_3.

In the internal granular layer (IGL) 9% of the cells were labelled (Fig. 2). There was an increase in the number of labelled cells on the third day. Between the 2rd and 5th day there was no change. On the 5th day a gradual decrease started and only

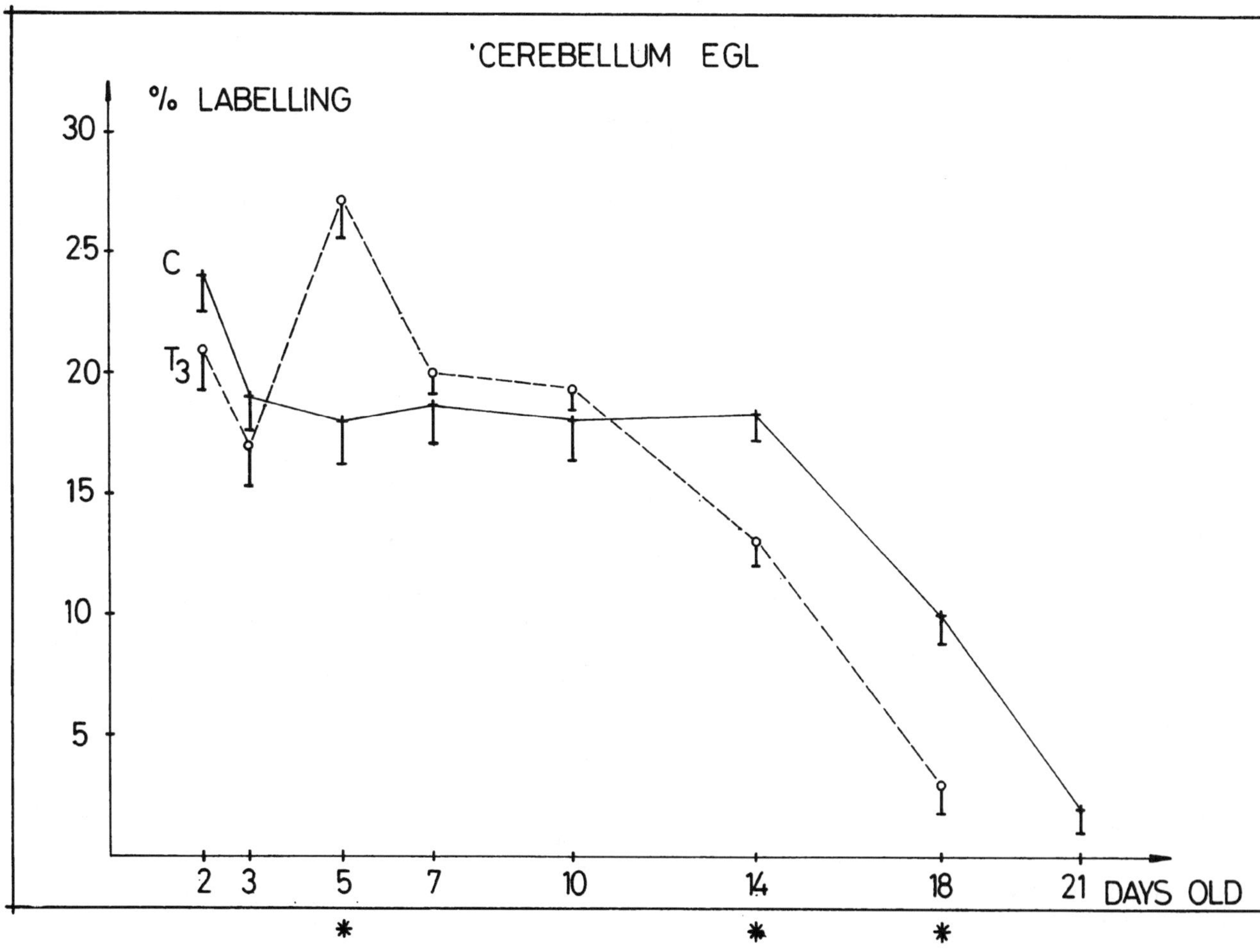

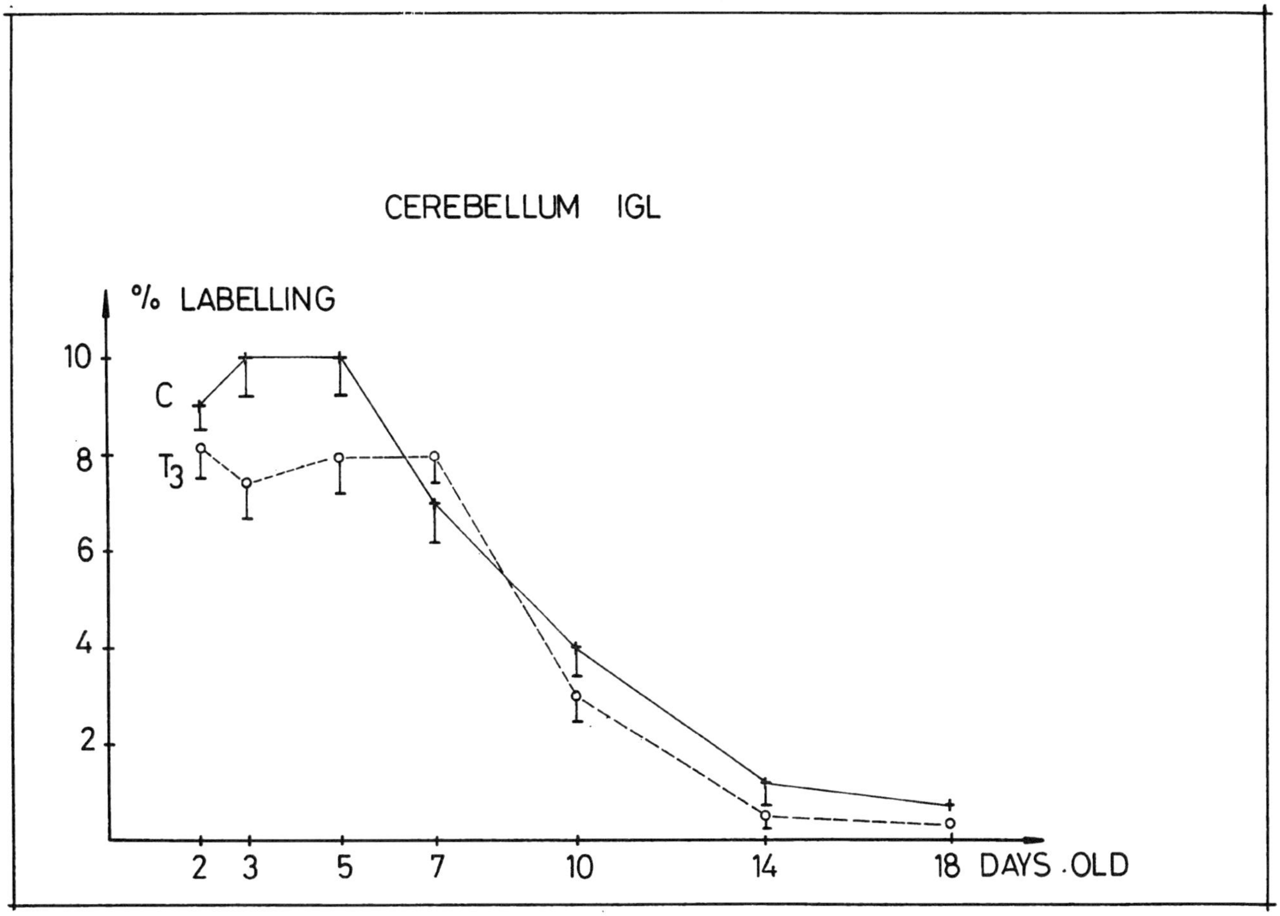

Fig. 3. Labelling indices in the IGL of the cerebellum.
1 hour labelling with 6-^{3}H-thymidine.
C: control
T_3: treated with triiodothyronine

0.5% of cells were labelled on the 18th day. There were no labelled cells on the 21st day. The T_3 treatment resulted in a temporary decrease from the 2nd to 3rd day. The labelling index was found higher on the 5th day and remained constant till the end of the first week.

On the 7th day a gradual decrease started and there were no labelled cells on the 21st day.

In the white matter of cerebellum 10% of the cells were labelled on the 3rd day (Fig. 4). This value increased till the end of first week and from the 7th to the 18th day it decreased to 1%. On the 21st day there were no labelled cells in this layer.

T_3 treatment caused a greater increase in the labelling than it was found in the control animals on the first week, but from the 10th day the number of labelled cells was found significantly smaller. On the 18th day there were no labelled cells in the T_3 treated animals.

Comparing the increase of the EGL in the control and T_3 treated animals it was found that the thickness of this cell layer increased till the 10th day and then decreased gradually. In the T_3 treated animals the thickness of EGL increased only till the 7th day. The EGL disappeared on the 24th day in the control and on the 21st day on the T_3 treated animals.

Telencephalon.

Numerous labelled cells were found in the dentate gyrus and in the periventricular germinal layer (PVGL) during the first three postnatal weeks (Fig. 5).

The number of labelled cells increased in the PVGL on the first week (Fig. 6). On the 7th day a decrease started. The layer disappeared about the end of the 3rd week.

The T_3 treatment resulted in a decrease between the 2nd and 3rd day. From the 3rd day the labelling index increased till 7th day and then it decreased again. There was a significant difference between the values of control and treated animals on the 3rd, 14th and 18th days. Many labelled cells were found in the granular and polymorph layers of the dentate gyrus (Fig. 7).

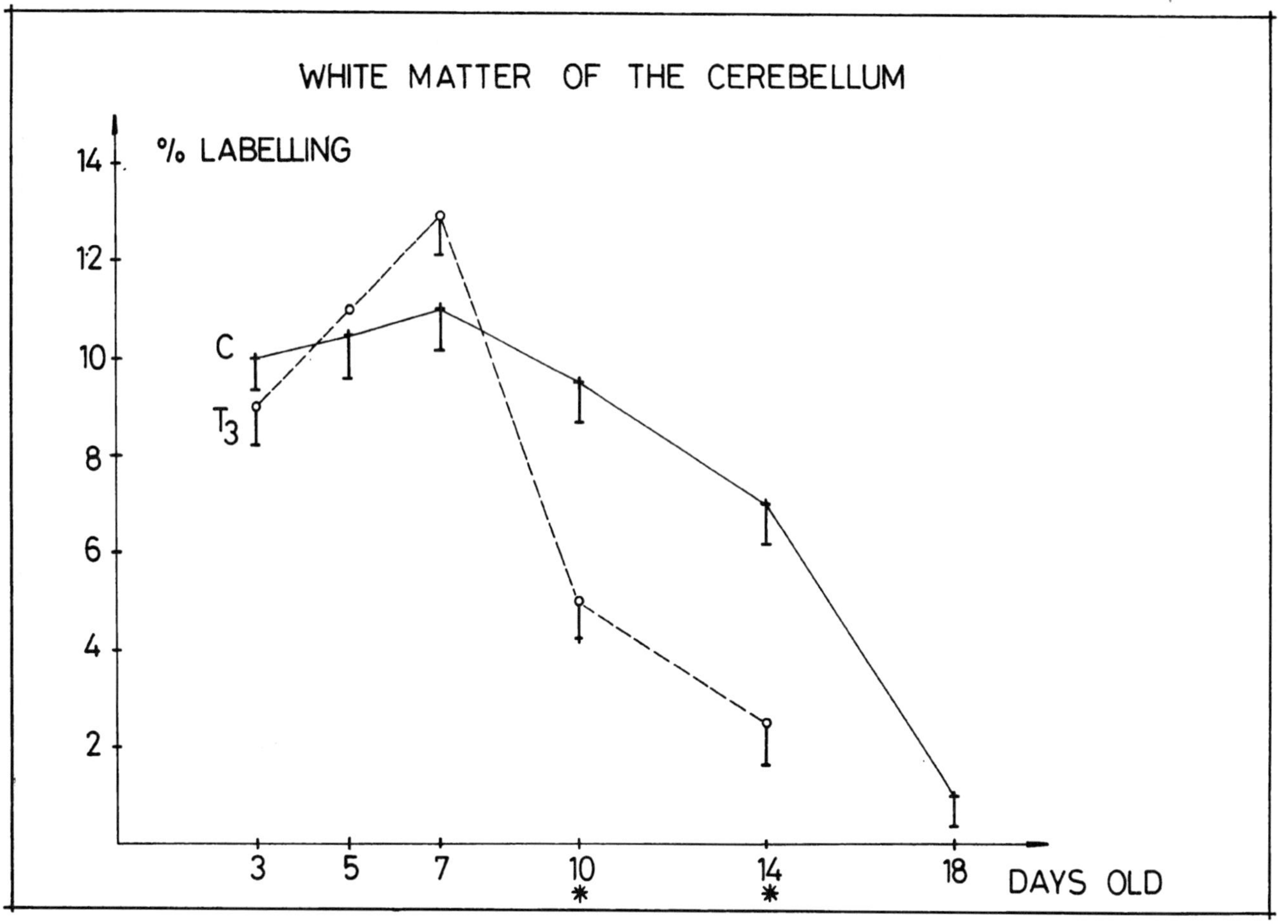
WHITE MATTER OF THE CEREBELLUM
% LABELLING
14
12
10
8
6
4
2
C
T3
3
5
7
10
*
14
*
18
DAYS OLD

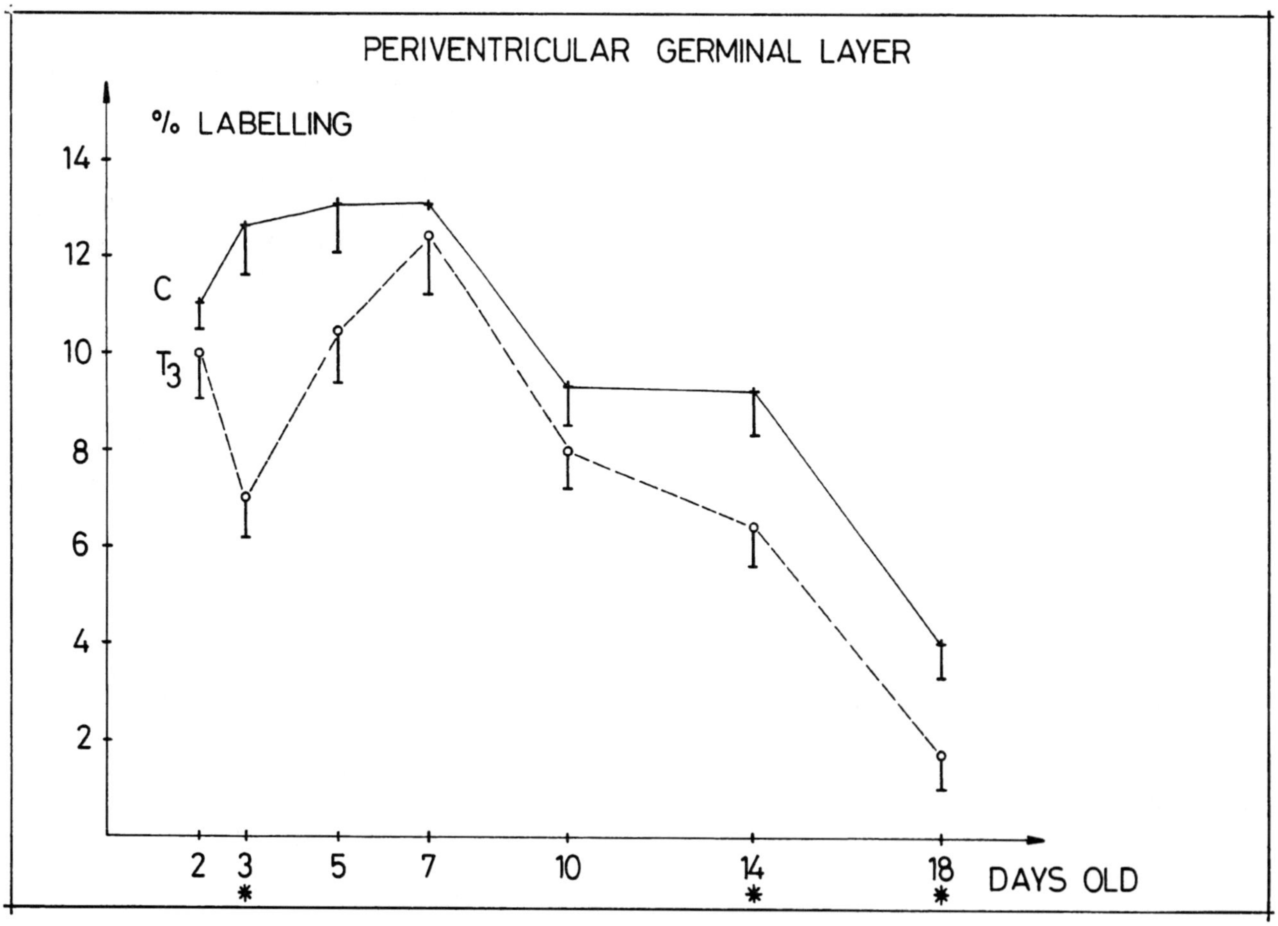
PERIVENTRICULAR GERMINAL LAYER
% LABELLING
14
12
10
8
6
4
2
C
T3
2
3
*
5
7
10
14
*
18
*
DAYS OLD

Six per cent of the cells were found labelled in the granular layer of dentate gyrus on the second day (Fig. 8). There was a considerable decrease from the 2nd to the 3rd and then a moderate increase till the 5th day. The labelling index decreased gradually from the 5th day and no labelled cells were found on the 21st day. The T_3 treatment caused no significant changes in labelling but several pycnotic nuclei were seen in this layer in treated animals from the 10th day.

Fig. 4. (S. 364, oben) Labelling in the white matter of the cerebellum.
1 hour labelling with 6-^{3}H-thymidine.
* ($p < 0.05$)
C: control
T_3: trented with triiodothyronine

Fig. 6. (S. 364, unten) Labelling indices in the periventricular germinal layer. 1 hour labelling with 6-^{3}H-thymidine
* ($p < 0.05$)
C: control
T_3: treated with triiodothyronine

Fig. 8. Labelling indices in the granular layer of the dentate gyrus. 1 hour labelling with 6-^{3}H-thymidine
C: control
T_3: treated with triiodothyronine

(Fig. 9) Ten per cent of the cells were labelled in the polymorph layer on the second day. There was a considerable decrease also in this layer from the 2nd to the 3rd day. Then the number of labelled cells increased till the 10th day. On this day a gradual decrease started.

The T_3 treatment caused significant decrease in labelling on the 10th and 14th day. There were no labelled cell on the 21st day.

Diencephalon

(Fig. 10 and 11) Labelled cells were seen in the hypothalamus and in the ependyma of the third ventricle during the first two weeks of life. The labelling indices were low (2–4%) and the T_3 treatment had no effect on the postnatal cell formation of the hypothalamus.

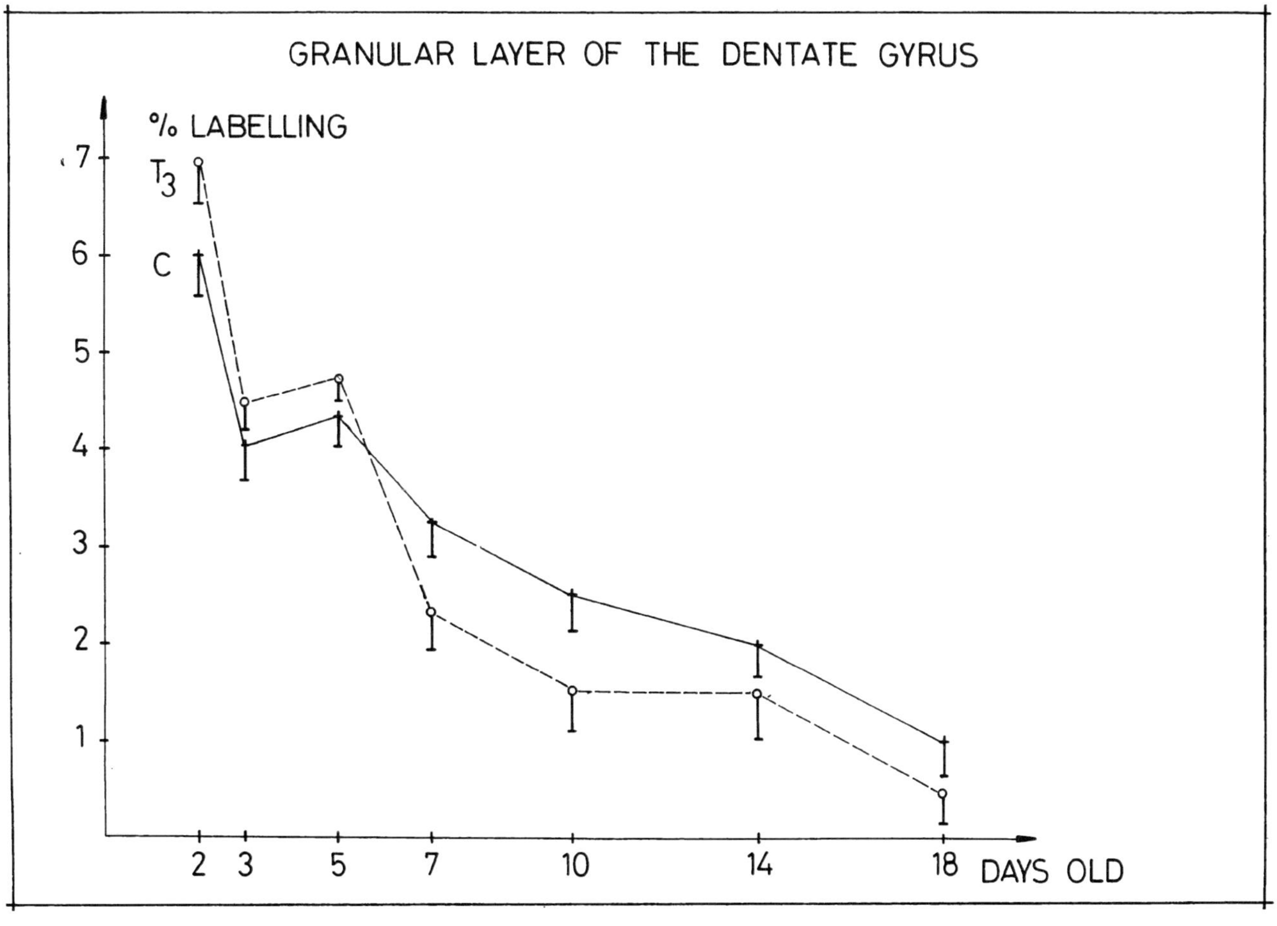

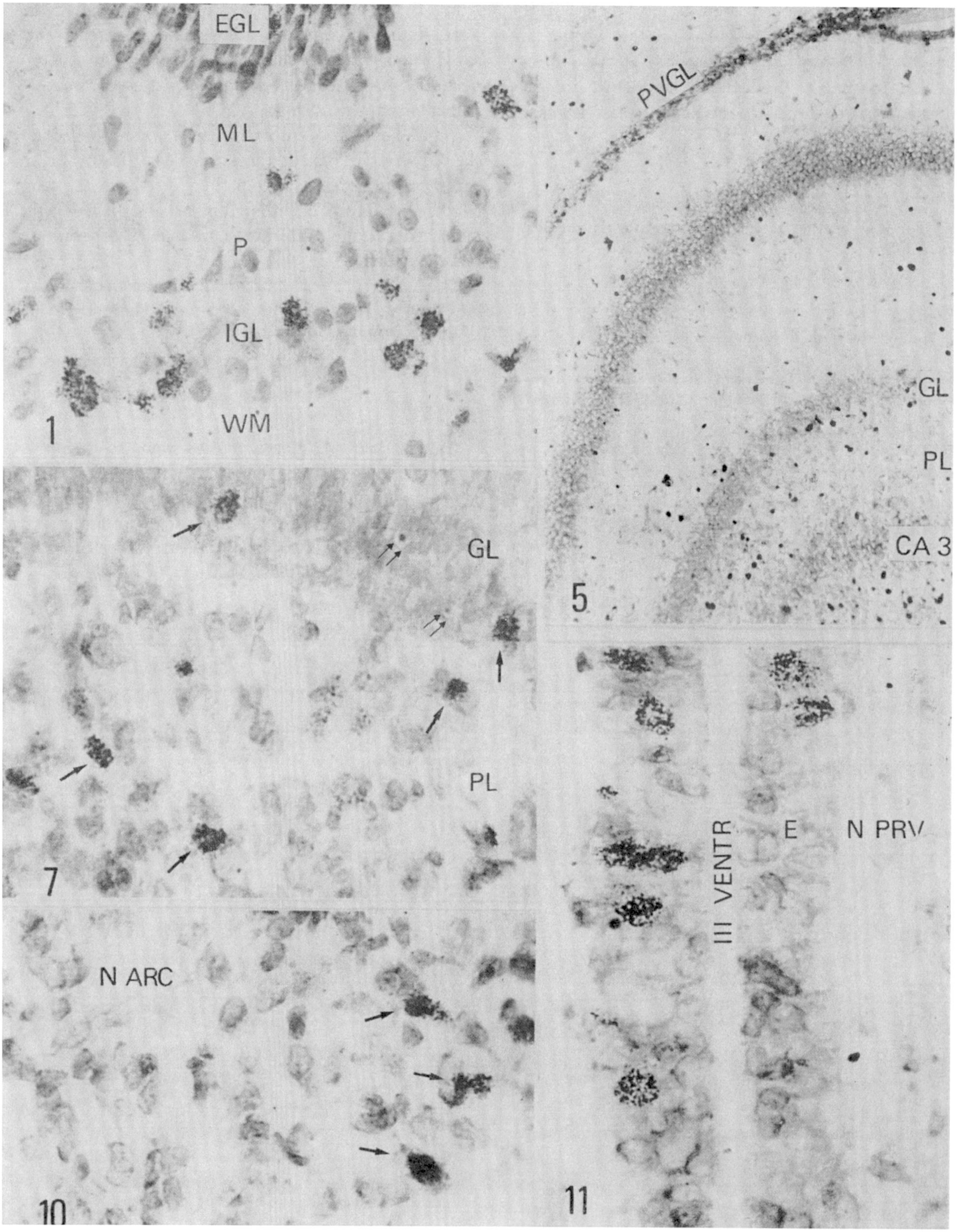
EGL
ML
P
IGL
WM
1
PVGL
GL
PL
CA 3
5
GL
PL
7
III VENTR
E
N PRV
N ARC
10
11

Fig. 1, 5, 7, 10, 11

Fig, 1. Labelled cells were found in all layers of the cerebellum.

5 days old control animal.

1 hour labelling with 6-^{3}H-thymidine 5 μ paraffine section. Stained with gallocyanine. × 500

EGL: external granular layer
ML: molecular layer
IGL: internal granular layer
P: Purkinje cell layer
WM: White matter of cerebellum

Fig. 5. Labelled cells in the dentate gyrus and in the periventricular germinal layer of a 5 days old control animal. 1 hour labelling with 6-^{3}H-thymidine. 5 μ paraffin section. Stained with Nissl method. × 80

PVGL: periventricular germinal layer
GL: granular layer
PL: polymorph layer
CA 3: Ammon's horn (area CA 3)

←

Fig. 7. Labelled cells (arrows) in the granular and polymorph layer of the dentate gyrus. Double arrows indicate pycnotic nuclei 10 days old T_3 treated animal 5 μ paraffin section. Stained with Nissl method. × 500

GL: granular layer
PL: polymorph layer

Fig. 10. Labelled cells (arrows) in the nucleus arcuatus of the hypothalamus. 1 hour labelling with 6-^{3}H-thymidine 5 days old control animal. 5 μ paraffin section. Stained with Nissl method × 500

N. ARC.: nucleus arcuatus

Fig. 11. Labelled cells in the ependyma of the third ventricle 5 days old control animal. 1 hour labelling with 6-^{3}H-thymidine.

5 μ paraffin section. Stained with Nissl method × 500

III. Ventr.: third ventricle

E: ependyma

N. PV.: nucleus periventricularis

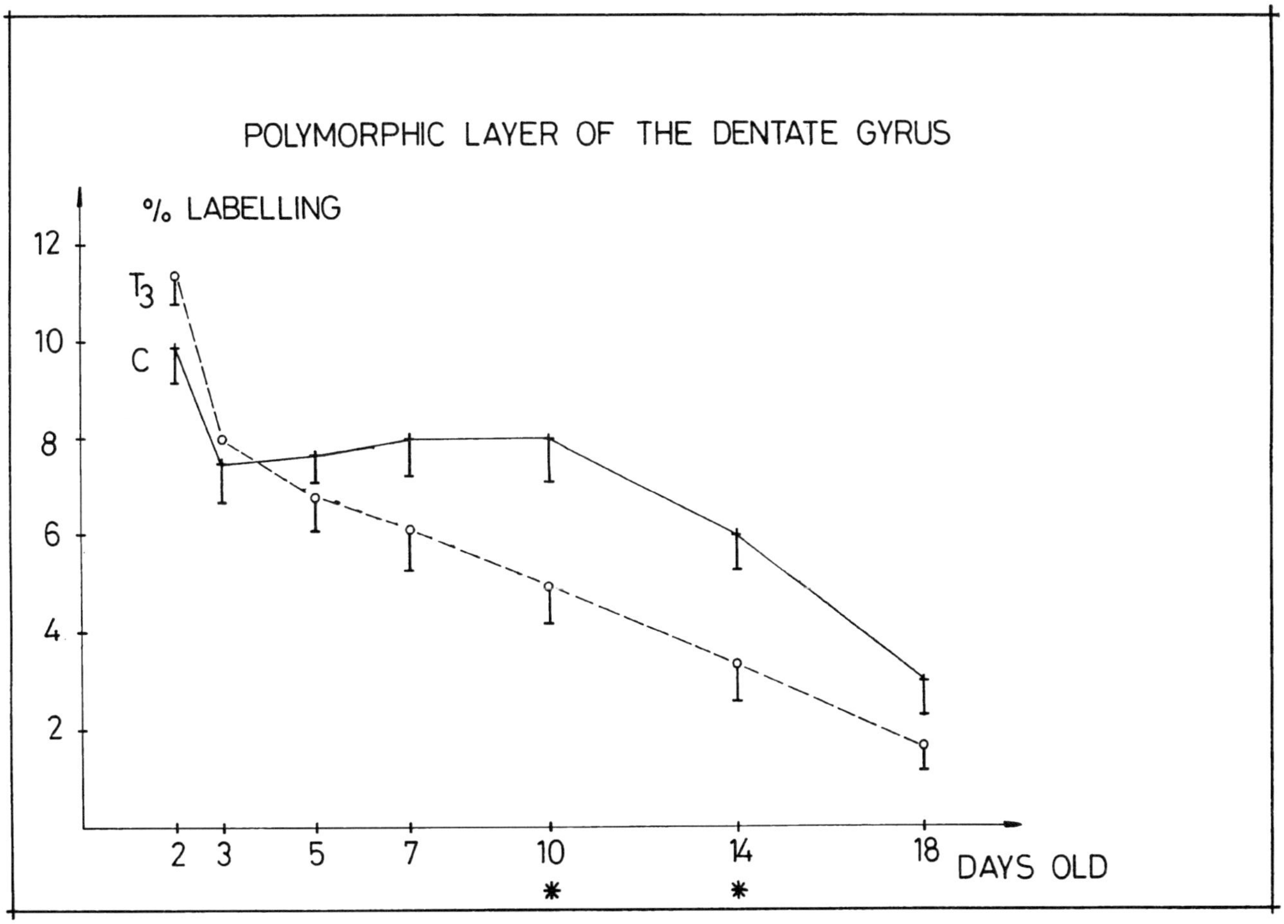

Fig. 9. Labelling indices in the polymorph layer of the dentate gyrus. 1 hour labelling with 6-^{3}H-thymidine.

* ($p < 0.05$) C: control T_3: treated with triiodothyronine

Discussion

It seems that daily doses of 5 μg/rat T_3 have different effects upon the various secondary germinal layers and have no effect at all on the postnatal cell formation of the hypothalamus.

In the periventricular germinal layer which produces microneurons and glial cells T_3 causes a decrease of cell formation.

In the EGL of the cerebellum (which produces only microneurons) as a result of daily T_3 treatment an enhancement of cell proliferation was observed during the first week followed by a decrease on the second week.

For the cerebellum similar results were reported by Gourdon et al. (1973) and Kovács et al. (1975) showing the increase of DNA content of the cerebellum on the first, and a decrease from the second week of life.

A substantial number of proliferating cells was observed in the IGL. This was also observed by several workers (Miake et al., 1961; Altman 1966, Altman et al., 1966; Fujita et al., 1966). The normal time course of cell formation and T_3 dependence of this process, has now been described.

It can be questioned that which cell type is formed within the IGL, as Purkinje and Golgi cells are already present at birth and granule cells are acquired through migration (Altman 1966, Rakic et al., 1970). Considering that 1 hour exposure period was used for the incorporation of thymidine into the cell nuclei, the migration of cells between the time of isotope injection and killing of the animals could be ruled out. Postmigratory granule cells are unlikely to divide because of their wel developed axonal system. In our opinion proliferating cells may correspond to the astroglia.

In the white matter of the cerebellum T_3 resulted in a marked decrease in the number of labelled cells from the 10th day. These are oligodendroglia cells. This might be the background of the disturbance of myelination observed following postnatal T_3 treatment (Balázs et al., 1971a, 1971b). We could observe a similar marked decrease of the number of labelled oligodendroglia cells in the corpus callosum from the 10th day of life (unpublished data) showing that the T_3 treatment influenced not only the myelination of the cerebellum but that of other brain regions too.

The daily T_3 treatment resulted in a decreasing labelled cell number in the dentate gyrus but there was no difference in cell number in adulthood between the treated and control animals (unpublished). In the hypothalamus the neurons are thought to be acquired mainly prenatally (Ifft, 1972) but in this work postnatal cell formation could be seen till the end of the second week. The postnatally formed cells may correspond to microneurons or to glia cells. The T_3 treatment had no effect on this postnatal cell formation.

In the present investigations it was found that the postnatal T_3 treatment acts differently on the cell formation in the secondary germinal layers. Similar results were reported by Lewis et al. (1977) using reserpine. They found differences in the effect of reserpine on the EGL of the cerebellum compared to that in the periventricular germinal layer of the forebrain.

The reason of divergent effects of the postnatal T_3 treatment might be the different cell (neuron and/or glial cell) production in the various germinal layers, but the intensity of cell division can also be responsible.

References

Altman, J., and G. D. Das: Autoradiographic and histological studies of postnatal neurogenesis. I. A longitudinal investigation of the kinetics, migration and transformation of cells incorporating tritiated thymidine in neonate rats, with special reference to postnatal neurogenesis in some brain regions. J. Comp. Neurol. **126**, 337–390 (1966).

Altman, J.: Autoradiographic and histological studies of postnatal neurogenesis. II. A longitudinal investigation of kinetics, migration and transformation of cells incorporating tritiated thymidine in infants rats, with special reference to postnatal neurogenesis in some brain regions. J. Comp. Neur. **128**, 431–474 (1966).

Balázs, R., S. Kovács, W. A. Cocks, A. L. Johnson and J. T. Eayrs: Effect of thyroid hormone on biochemical maturation of rat brain: Postnatal cell formation. Brain Research **25**, 555–570 (1971).

Balázs, R., W. L. Brookshank, A. J. Patel, A. L. Johnson and D. A. Wilson: Incorporation of S^{35}-sulphate into brain constituens during development and the effects of thyroid hormone on myelinisation. Brain Research **30**, 273–293 (1971).

Fujita, S., M. Snimada and T. Nakamuro: H_3-thymidine autoradiographic studies on the cell proliferation and differentiation in the external and internal granular layers of the mouse cerebellum. J. comp. Neurol. **128**, 191–208 (1966).

Gourdon, J., J. Clos, C. Coste, J. Dainat and J. Legraund: Comparative effects of hypothyroidism, hyperthyroidism and undernutrition on the protein and nucleic acid contents of the cerebellum in the young rat. J. Neurochemistry **21**, 861–871 (1973).

Hamburg, M.: An analysis of the action of thyroid hormone on development based on in vivo and in vitro studies. Gen. and Comp. Endocrinol. **10**, 198–213 (1968).

Ifft, J. D.: An autoradiographic study of the time of final division of neurons in rat hypothalamic nuclei. J. Comp. Neurol. **144**, 193–205 (1972).

Kovács, S.: The role of thyroid and adrenocortical hormones in the biochemical maturation of the rat brain, in Hormones and Brain Function. Plenum Press pp. 53 to 67 (1973).

Kovacs, S., and L. Seress: The effect of neonatal triiodothyronine treatment on the development of the rat brain. XLI. Annual Meeting of Hungarian Physiological Society (1975) (in press).

LEWIS, P. D., A. J. PATEL, G. BÉNDEK and R. BALÁZS: Effect of reserpine on cell proliferation in the developing rat brain: A quantitative histological study. Brain Research **129**, 299—308 (1977).

MIALE, J. L., and R. L. SIDMAN: An autoradiographic analysis of histogenesis in the mouse cerebellum. Exp. Neurol. **4**, 277—296 (1961).

NICHOLSON, J. L., and J. ALTMAN: The effect of early hypo- and hyperthyroidism on the development of rat cerebellar cortex. I. Cell proliferation and differentiation. Brain Research **44**, 13—13 (1972).

RAKIC, P., and R. L. SIDMAN: Histogenesis of cortical layer; in human cerebellum, particularly the lamina dissecans. J. Comp. Neurol. **139**, 473—500 (1970).

Author's address:

Dr. László SERESS
1st Department of Anatomy
Nagasaki University Medical School
Sakamoto cho 12—4
Nagasaki 852 — Japan

Buchbesprechung

SUCHENWIRTH, R.:

Taschenbuch der klinischen Neurologie

2. Aufl., 119 Abb., 221 S., 12 × 19 cm. Gustav-Fischer-Verlag Stuttgart. Brosch. DM 16,80. ISBN 3-437-00217-1

Lehrbuchartige Darstellungen der klinischen Neurologie gibt es in großer Zahl und in allen Größenordnungen. Die Vielfalt der für Medizinstudenten geschriebenen Einführungen wird u. a. durch das zu Recht bestehende Bedürfnis nach Berücksichtigung unterschiedlicher Lehrmeinungen hervorgerufen. Grundsätzliche Veränderungen der äußeren Form und der Anordnung der Stoffvermittlung gegenüber den bisherigen Gepflogenheiten sind selten. Eine solche Ausnahme bildet dieses Taschenbuch, dessen 1. Auflage sehr schnell hohe Anerkennung fand. Auf den ersten Blick ein Wagnis, die klinische Neurologie auf 200 kleinformatigen Seiten, die zum großen Teil durch Abbildungen gefüllt sind, mit ganz knappem Text abzuhandeln, zeigen inzwischen Examenerfahrungen doch, daß sich das Büchlein gerade für die Prüfungsvorbereitung bewährt hat. Freilich ist das Buch allein nicht dafür ausreichend, aber bei kontinuierlicher Unterrichtsteilnahme oder/und ergänzendem Lehrbuchstudium erleichtert es dem Studenten Verständnis und Einprägung. Dies wird vor allem durch die aussagekräftigen, teils farbigen graphischen Darstellungen (D. ANDRIAN) erreicht, die nur eines knappen ergänzenden Textes bedürfen. Didaktisch besonders effektiv sind die Rückenmarksquerschnitte sowie die bildliche Veranschaulichung der peripheren Nervenläsionen.

Damit erweist sich erneut die enorme Bedeutung der neuroanatomischen Grundlagen unseres Faches, ohne die einfach der Zugang zu neurologischem Denken und Verstehen versperrt ist. Die nicht minder große Wichtigkeit funktioneller, physiologischer und biochemischer Zusammenhänge soll damit nicht eingeschränkt werden.

Im Vergleich beider Auflagen ist festzustellen, daß sich der Umfang nur um knapp 20 Seiten, die Zahl der Abbildungen um 9 erhöht hat. Die Einteilung in topische Neurologie und neurologische Nosologie wurde konsequent beibehalten. Dabei hat der erstgenannte Teil nur wenige Veränderungen bzw. Ergänzungen im Text erfahren. Im zweiten Teil wurden zwei wichtige (in der 1. Auflage mit Recht vermißte) Querschnittsbilder der Medulla oblongata und des Pons aufgenommen. Auch die Übernahme der schematischen Abbildung der Äste der A. cerebri media nach FOIX u. LEVY und die Einfügung einer Tabelle der zerebralen Durchblutungsstörungen sind zu begrüßen. Das Tumorkapitel wurde textlich und durch Abbildungen ergänzt. Unter der Überschrift „der Nerv in der Falle" werden die wichtigsten Kompressionssyndrome aufgeführt. Statistische Angaben über die Häufigkeit neurologischer Krankheiten, Zusammenstellungen der wichtigsten heriditären Krankheitsgruppen und eine graphische Übersicht der Nervenleitgeschwindigkeiten unter normalen und pathologischen Bedingungen wurden am Ende des Buches angefügt.

Das Taschenbuch hat sich schon seinen festen Platz gesichert. Es ergänzt das Schrifttum in origineller Weise und ist von hohem didaktischen Wert. Es empfiehlt sich vor allem als zusätzliche Lernhilfe und zur schnellen Auffrischung der Kenntnisse.

H. A. F. SCHULZE, Berlin

HEILIGENBERG, W.: Principles of Electrolocation and Jamming Aviodance in Electric Fish.

A Neuroethological Approach.
Springer Verlag Berlin—Heidelberg—New York.
85 Seiten, 58 Abbildungen, 1 Tabelle US-Dollar 9,50.

Das Buch eröffnet eine geplante Serie von Abhandlungen von Gehirnfunktionen, wobei die Betonung auf Aspekten der Hirnfunktion in bezug auf ihre Verhaltensäußerungen liegt, nicht aber vordergründig Entwicklungsprobleme behandelt. Während komplizierte Verhaltensakte nur selten bis auf das neuronale Niveau zurückgeführt werden können (wenn man von Korrelationen absieht), macht laut den Ausführungen des Verfassers der elektrische Fisch insofern eine Ausnahme, als „Reizaufnahme und -bewertung sowie Steuerung spezifischer Verhaltensantworten direkt auf eine fast kontinuierliche Kette von beobachtbaren neuronalen Ereignissen" bezogen werden können. Zunächst wird auf 7 Seiten eine knappe Übersicht zur allgemeinen Physiologie des elektrischen Organs, der Elektrorezeptoren, der Taxonomie elektrolokalisierender Fische und zur Neuroanatomie (Hirnstrukturen der Mormyriformen und der Gymnotoiden) gegeben. Der Hauptteil des Buches ist dem Mechanismus der Elektrolokation gewidmet. Nachdem räumliche Aspekte der Elektrolokation, d. h., der Lokalisation von Objekten die das vom elektrischen Organ bei dessen Entladungen aufgebaute elektrische Feld stören und damit zu einer Veränderung der Rezeption des Feldes durch die im Seitenlinienorgan mitbefindlichen Elektrorezeptoren führen, behandelt worden sind, wird auf die Antwortcharakteristika und die zentralen Projektionen der Elektrorezeptoren eingegangen. Die Unterschiede der Rezeptorantwort bei Puls-Species (Entladungen des elektrischen Organs mit einer Sequenz ähnlich jener von Nervenzellen) und bei Wellen-Species (Entladungen höherer Frequenz) werden herausgestellt. In einem weiteren Kapitel wird die zentrale Verarbeitung der von den Rezeptoren stammenden Eingänge zur „elektrischen Abbildung" in bezug auf Formen, Bewegungen der Objekte usw. behandelt. Verhaltensstudien werden anschließend berichtet, darunter die sogenannte jamming avoidance, d. h. z. B., daß Wellen-Species wie Eigenmannia und Gymnarchus bei Vorhandensein sinosoidaler Signale nahe der Entladungsfrequenz ihres eigenen elektrischen Organs diese aus der Frequenzzone des fremden Signals verlagern, um so ihre Fähigkeit zur Elektrolokation aufrechtzuerhalten. Zum Schluß werden Hypothesen über die Evolution von Wellen-Species der elektrischen Fische und der damit verbundenen Vorteile in der Elektrolokation erörtert.

W. RÜDIGER, Berlin

J. Hirnforsch. 19 (1978) 405–414

Aus dem Zoologischen Institut der Universität Basel
(Vorsteher: Prof. Dr. H. NÜESCH und Prof. Dr. W. STINGELIN)

Untersuchungen zur Embryonalentwicklung des Gehirns von Scyliorhinus canicula /L./ III. Das optische System und angrenzende Nuclei im mesencephalen Tegmentum

Von Hans-Peter FARNER[1])

Mit 5 Abbildungen

(Eingegangen am 27. Juli 1977)

Kurztitel: Embryonalentwicklung des Gehirns von *Scyliorhinus c.*

Summary: The ontogenesis of the optic system in the brain has been studied with reference to the general biology of the embryo and to the cytoarchitecture of the optic tectum.

1. Sharks vary in terms of retinal structure (FRANZ 1931). *Scyliorhinus* shows densely packed and large cellular layers alternating with thin fiber layers.
2. The optic tracts, which are not excessively large, cross in the optic chiasm by approximately 95%. The optic fibers, forming a medial and a lateral tract, enter the tectum in zones 9 and 11.
3. The eye muscle nuclei do not differ from the general vertebrate pattern. The oculomotor nucleus represents the largest portion; the abducens nucleus is small.
4. The nucleus interstitialis of the fasciculus longitudinalis medialis is remarkably developed and consists of large neurons. Together with the interpreduncula nucleus it receives fibers of fasciculus retroflexus.

Zusammenfassung: Die vorliegenden Untersuchungen über die embryonale Entwicklung von Kernarealen und Faserzügen am optischen System bestätigen Zusammenhänge zwischen der Lebensweise des Tieres einerseits und der Cytoarchitektonik des Tectum opticum andererseits .

1. Die Ausbildung der Retina ist bei verschiedenen Haiarten sehr unterschiedlich (FRANZ 1931). Der Aufbau der Retina zeigt sehr dichte, mächtige Zellagen, neben eher gering ausgebildeten Faserlaminae.
2. Der Schichtdicke retinaler Fasern im Auge entsprechend, ist der Tractus opticus wenig mächtig. In einzelne Kompartimente gegliedert, zieht er zum Chiasma opticum, wo sich die Fasern zu über 95% kreuzen. Proximal des Chiama spaltet sich der Tractus opticus in einen Tractus marginalis medialis, dessen Fasern sukzessive in Zone 9, vorwiegend in Zone 11 einstrahlen und einen Tractus marginalis lateralis, dessen Fasern sich fächerförmig über das ventrale Sehhügelgebiet erstrecken und in Zone 9 einziehen.
3. Unter den Augenmuskelkernen ist der Nucleus oculomotorius der auffälligste. Kaudal grenzt der Nucleus trochlearis an, beide umfassen und durchdringen von dorsal her den Fasciculus longitudinalis medialis. Aus wenigen Zellen bestehend und einen entsprechend dünnkalibrigen Nervus bildend, liegt der Nucleus abducens in der Medulla oblongata. Ontogenetisch erscheint der Abducenskern retardiert.
4. Der auffälligste Kern mit seinen Riesenneuronen ist der Nucleus interstitialis des Fasciculus longitudinalis medialis. Seine embryonale Entwicklung setzt sehr früh ein; die stark meylinisierten Axone erscheinen bereits in Em 13. Zusammen mit dem Ganglion interpedunculare können die beiden Kernareale als Endstellen des Fasciculus retroflexus bezeichnet werden.

Einleitung

Ist das optische System Gegenstand einer Untersuchung, so muß, aus Gründen der Klarheit eine Begrenzung erfolgen. Für die vorliegende Arbeit werden Retina, Tractus opticus und dessen Verlauf durch das Zwischenhirn, sowie die Nuclei der Augenmuskelnerven in Betracht gezogen. Das Tectum opticum und dessen Ontogenese wurde im vorangehenden zweiten Teil behandelt. Aus didaktischen Gründen wurde, der bereits in der Medulla oblongata liegende Nucleus abducens mit den übrigen Augenmuskelkernen im Abschnitt Kerene des mesencephalen Tegmentums untersucht. Als nicht direkt zum optischen System gehörend, gelangten tegmentale Kernareale zur Untersuchung, die mit Nuclei des Systems Faserkontakte aufweisen.

Bereits der Bau der Retina gibt innerhalb der

[1] Die vorliegende Arbeit entstand unter der Leitung von Herr P. D. Dr. D. G. SENN für dessen fruchtbare Anregungen und Diskussionen ich herzlich danke. Für das stets gezeigte Interesse an dieser Problemstellung danke ich Herrn Dr. W. STINGILIN. Den Herren Dres. Jürg MARTY und Dieter FRÖSCH (Lobaratoires Arago, Banylus sur Mer) verdanke ich die Anlieferung der Eier. Ein besonderer Dank gilt dem Gönner der Ichthyologische Forschung Herrn Dr. A. SIGG (Hergiswil).

Selachii zu Diskussionen Anlaß. Der nicht einheitliche Bau der Retina innerhalb der Plagiostomen wird noch unübersichtlicher durch die Tatsache, daß verschiedene Autoren bei artgleichen Tieren entweder Stäbchen und Zapfen oder nur Stäbchen beschreiben. Bei weniger gut differenzierter Retina, wie sie bei Scyliorhinus canicula vorliegt, ist eine Entscheidung schwierig.

Faseranatomisch müßten Verbindungen von optischen Nuclei und Ganglien zum Telencephalon noch untersucht werden. Ebenfalls einer weiteren Arbeit sei es überlassen, die Arten der Kontaktnahme einzelner Kern- und Fasergebiete zu studieren.

Material und Methoden

Die in der vorliegenden Arbeit zur Untersuchung verwandten Embryonen wurden in künstlich angesetztem Meerwasser von 1,026 Dichte herangezogen. Die vom „Laboratoire Arago" in Banyuls sur Mer (52 Stück) und von der Stazzione Zoologica in Neapel (48 Stück) angelieferten Eier wurden in einem 150 Liter Aquaroium bei 15,5 ± 0,5° Celsius und bei pH 8 bis 8,3 bis zum gewünschten Entwicklungszustand gehalten. Aufgrund stetiger Kontrollen im Durchlicht wurden die Embryonen nach Entwicklungsgrad gestaffelt jeweils der Eischale entnommen und fixiert. Die exakte Datierung ist bei *Scyliorhinus canicula* im voraus nicht möglich, da zum Zeitpunkt der Eiablage, der seinerseits unbekannt ist, der Entwicklungszustand der Keimscheibe individuell variiert. Die Datierung einschließlich der Vermessung der Embryonen wurde in der vorangehenden Arbeit (Farner 1977) publiziert.

Die Tiere wurden in den Eihüllen beobachtet, mit Wendling- und Scammon-Angaben verglichen, im gewünschten Stadium photographiert, mit MS 222 narkotisiert und in AFE (90 ml Alkohol 80%, 5 ml Formol 40%, 5 ml Eisessig) fixiert. Die Embryonen mit den fortlaufenden Nummern 1 bis 42 wurden den Eihüllen entnommen. Em 43—52 sind Schlüpflinge, deren durchschnittliche Entwicklungszeit 157 Tage betrug. Das adulte Gehirn wurde einem Tier von 45 cm Länge entnommen, das zur Sektion mit MS 222 narkotisiert wurde. Die ganzen Köpfe der Embryonen Em 21—45 wurden von 18 bis 48 Stunden in Ttiriplex (ADTE = Aethylendiamintetraessigsäure) entkalkt. Alle Präparate in Paraplast eingebettet, in Schnittserien verschiedener Schnittrichtungen und -dicken verarbeitet und mit der Neurofibrillendarstellung nach Bodian mit Albumosesilber imprägniert und anschließend mit Kresylviolett (Senn 1966) gefärbt.

Bau und Entwicklung der Retina

Innerhalb der Selachier weicht der Bau der Retina wesentlich vom allgemeinen Wirbeltierschema ab. Insbesondere variieren Literaturangaben über die Zusammensetzung der Sehzellenschicht. V. Franz (1931) hat bezüglich des Vorhandenseins von Zapfen bei Scyliorhinus canicula die Angaben von acht Autoren verglichen. Drei haben Stäbchen und Zapfen und fünf nur Stäbchen festgestellt. Die vorliegenden Präparate ergeben nirgends Anhaltspunkte für das Vorhandensein von Zapfen.

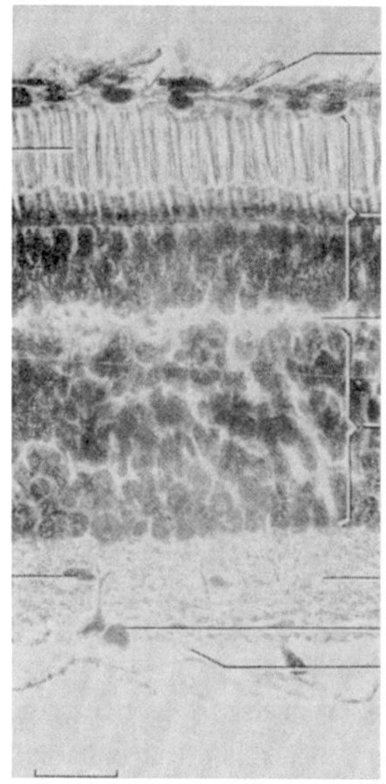

Abb. 1. Querschnitt durch die Retina eines Schlüpflings. Die sehr lockere Ganglienzellschicht bildet eine schmächtige Schicht von Opticusaxonen.
Aufnahme mit Phasenkontrastoptik. (Maßeinheit = 25 μ)

Die Schichtung der Retina entspricht in den Proportionen durchaus dem allgemeinen Selachiertyp. (Abb. 1) Dem Pigmentepithel folgt ein langausgezogener Stäbchenraum. Der äußeren Körnerschicht folgt eine wenig mächtige äußere plexiforme Schicht, die vereinzelt von Zellen durchsetzt ist. Die innere Körnerschicht besteht aus zwei Teilschichten, die äußeren Zweidrittel bilden die eigentliche innere Körnerschicht; wobei das innere Drittel aus amakrinen Zellen und den Perikaryen der Müllerschen Zellen besteht. Die innere plexiforme Schicht führt Neurone der Ganglienschicht; sie ist etwa gleich mächtig wie der Stäbchensaum. Die Hauptmenge der Ganglienzellen bildet eine einschichtige innerste Zone, die lediglich von der dünnen Schicht retinaler Fasern überlagert ist. Es besteht ein starkes Übergewicht im Verhältnis zwischen Sehzellen und Ganglienzellen, d. h. auf mehrere Stäbchenzellen entfällt nur eine Ganglienzelle. Im Vergleich mit Netzhautstrukturen anderer Wirbeltiere ist diese zahlenmäßige Diskrepanz dahingehend zu deuten, daß die im schwachen Dämmerlicht noch funktionsfähige Retina, bedingt durch die große Sehzellenzahl einen Verstärkereffekt aufweist.

In Em 1 ist die Augenblase bereits ausgestülpt, die Linsenplakode erkennbar. Zwischen Em 3 und Em 4 stülpt sich die Retina ein und wird verdickt; die Linse wird abgeschnürt. Die in Em 4 noch verstreuten Mitosezellen sind in Em 5 streng auf die Zone des Ventrikelspaltes reduziert, wo sich gleichsam die

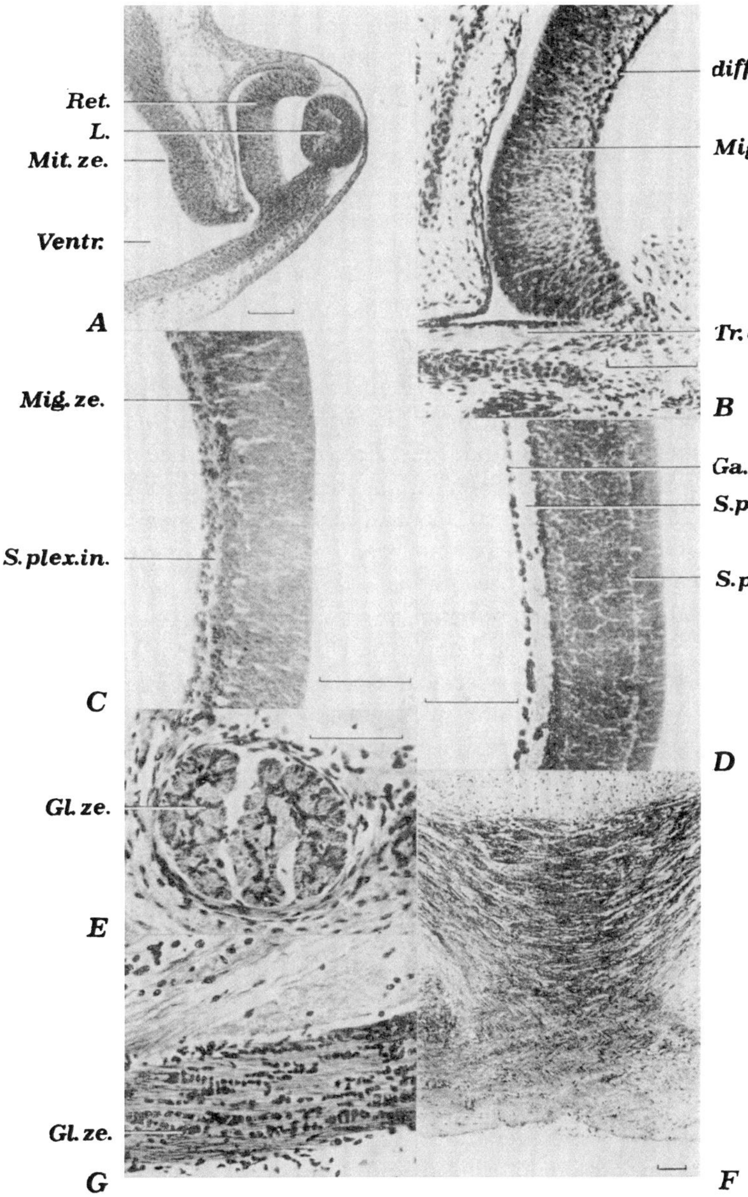

Abb. 2. Entwicklung der Retina und Aufbau des Tractus opticus. A: Querschnitt durch den Augenbecher von Em 5; B bis D: Retinaquerschnitte B von Em 11, C von Em 16 (die Mitosezellen befinden sich ortsgemäß in periventrikulärer Position), D von Em 23. E: Querschnitt durch den distalen Tractus opticus; F: Querschnitt durch das Chiasma opticum mit gegenseitiger portionenweiser Durchdringung der Faserbündel; G: Längsschnitt durch den proximalen Tractus opticus. (Maßeinheit = 100 μ)

retinale Matrix bildet. Bis Em 9 ist eine konstante Zellzunahme zu beobachten; darnach setzen die Differenzierung der Ganglienzellen und das Faserwachstum ein. (Abb. 2B) Im Em 11 besteht bereits ein Teil des Tractus opticus und des Chiasma. Mit Em 14 setzt die Differenzierung der Sehzellen ein. Als dünne Faserlamina bildet sich die innere plexiforme Schicht, der eine Lage amakriner Zellen in Em 16 angrenzt. Im späteren Sehzellenbereich sind die Mitosen sehr häufig. (Abb. 2c)

Die innere plexiforme Schicht nimmt rasch an Mächtigkeit zu und erreicht in Em 30 die Dicke des Schlüpfzustandes. Die äußere plexiforme Schicht tritt in Em 23 erstmals auf, erreicht bis Em.28 ihre endgültige, nur geringe Dicke (Abb. 2D). Die äußere Körnerschicht weist in Em 20 erste Differenzierungen auf, ist in Em 23 bereits in die typisch länglichen Zellen umgestaltet. Der Stäbchensaum ist deutlich erkennbar, und wird bis Em 28 stark erweitert. In Em 33 ist die Retina durchwegs dem Zustand des Schlüpflings sehr nahe. Mit der Volumenvergrößerung des Auges geht einerseits eine Auflockerung der Zellareale und der Fasermassen, andererseits eine Zunahme der Fasern einher.

Der Tractus opticus

Der verhältnismäßig dünne Tractus opticus zieht vom Auge rostromedial zum Chiasma opticum. Im distalen Teil ist der Faserzug in drei Haupt- und über 35 Nebenbündel gegliedert, wobei die Gliazellen in Reihen geordnet, die Bündelgliederung bewirken. (Abb. 2E) Proximal steigt die Dichte der Gliazellen, die Faserverteilung im Querschnitt erscheint diffus und der Tractus ist dorso-ventral abgeplattet.

Im Chiasma opticum liegt eine gestaffelte Faserschichtung derart vor, daß einer Faserlage des linken Auges eine solche des rechten folgt (Abb. 2F).

In Em 5 wachsen erste Fasern des Tractus opticus aus, also bevor sich die Ganglienzellen sichtbar differenzieren. Bis Em 9 wird die Anzahl kaum erhöht, dafür stärker myelinisiert. Die Fasern ziehen, ventral dem Augenstiel anliegend, zur Basis des Zwischenhirns. Die Gliazellen des Tractus opticus bilden noch in Em 13 einen zylinderförmigen Mantel um die optischen Fasern. Mit Em 16 hat der Tractus bezüglich Struktur und Bündelung den Stand des Schlüpflings erreicht, so daß sich weitere Veränderungen auf Faserzunahme und Längenwachstum beschränken. Proximal des Chiasma — opticum zieht der Tractus opticus marginalis in Bündeln dorsalwärts, wobei sich vereinzelt ganze Faserzüge derart kreuzen, daß marginale ventrikelwärts umbiegen und innere gleichzeitig zur Peripherie ziehen. Dorsal schließen erste Neurone des Nucleus geniculatus laterals an, die im ausgewachsenen Tier sehr locker gestreut liegen, und insgesamt einen kleinen Kern bilden. Tectumwärts trennt sich der Tractus opticus in zwei Teile; einen kompakteren Tractus marginalis medialis und einen auf den peripheren Zonen ausgebreiteten Tractus marginalis lateralis. Vor der Trennung überziehen retinale Faserbündel die beiden Tectumshälften in sphärischer Fächerform. Dorsocaudal des Nucleus geniculatus lateralis folgt der größere Nucleus geniculatus praetectalis, der lediglich zu lateralen optischen Bahnen Kontakt hat. Über die gesamte Länge der Sehhügel ziehen die Fasern des Tractus marginalis lateralis in die Ramonsche Zone 9 ins Tectum ein und bilden die tangentialen Axone dieser Schicht.

Die Fasern des Tractus marginalis medialis ziehen als bilaterale mediale Bündel im dorsalen Tectum kaudalwärts, wobei in Zone 9 vor allem auch in Zone 11 Fasern abzweigen, so daß sich die beiden Faserzüge nach kaudal erschöpfen.

Kerne im mesencephalen Tegmentum

Zellenanhäufungen im Tegmentum lassen sich in verschiedene Kernareale gruppieren. Der augenfälligste Kern, bezüglich Zellgröße ist der Nucleus interstitialis des Fasciculus longitudinalis medialis mit den großen polygonalen Zellen und den stark myelinisierten Axonen. Medio-dorsal, sich nach kaudal erstreckend, folgt der Nucleus oculomotorius, davon befindet sich der Nucleus trochlearis, die beide dem periventriculären Grau angehören. Latero-caudal des Nucleus interstitialis des F. l. m. liegt der Nucleus ruber. Auffallend und ausgeprägt zeigt sich in der Ventromedianebene der Nucleus interpeduncularis. Weit kaudal im Bereich der Medulla oblongata liegt, ebenfalls im periventriculären Grau, der Nucleus abducens, der als Augenmuskelkern hier einbezogen wird. (Abb. 3)

A) Nucleus oculomotorius

Unter den drei Augenmuskelkernen ist der Nucleus oculomotorius, wie allgemein bei Wirbeltieren stärker entwickelt und differenzierter als die Nuclei trochlearis und abducens. Er formiert sich medio-dorsal des Nucleus interstitialis des Fasciculus longitudinalis medialis und kaudalwärts liegt er dem Fasciculus an. Alle Zellen liegen im Stratum periventriculare und gehören, wenn man dorsalwärts die Zonen des Tectum opticum in die Betrachtungen einbezieht, den Zonen 3 und 5 an. Die multipolaren Zellen weisen etwa die vierfache Größe der schichtgleichen Zellen des Tectums auf. Die ventralwärts ziehenden Fasern gliedern den Fasciculus longitudinalis medialis in eine Anzahl Teilbündel, wobei andererseits die Fasern des Nervus oculomotorius bündelweise nach ventral ziehen und sich ventral des kaudalen Kerngebietes zur Oculomotoriuswurzel formieren und das Tegmentum verlassen. Im kaudalen Kerngebiet kreuzen sich einige Faserzüge des Nervus III. Auch sind direkte Verbindungen des Nucleus III mit dem Fasciculus longitudinalis medialis sehr naheliegend. Fasern der Zone 6 ziehen ventralwärts in den Kern ein, wobei diese sich teilweise ventromedian kreuzen und zum gegenüberliegenden Nucleus ziehen.

Ontogenetisch werden erste Neurone des Nucleus oculomotorius in Em 11 differenziert und sind bereits aus der Proliferationszone migriert; so daß die noch ventrikelwärts liegenden indifferenten Zellen das Kerngebiet um- und durchwandern müssen, d. h. die Zellen des Nucleus oculomotorius werden gleichsam überholt. Auch sind in Em 11 wenig myelinisierte Axone ausgebildet, die außerhalb des Gehirns bereits von Gliazellen begleitet sind, und in Em 13 eindeutig den Nervus oculomotorius bilden, der bis zum Auge auszumachen ist. Bis Em 16 wird der aus wenigen Zellen bestehende Nucleus durch die auswachsenden Fasern des Fasciculus longitudinalis medialis in einzelne Teile gegliedert, so daß die gegenseitige Faserdurchdringung aktiv vom Fasciculus longitudinalis medialis ausgeht. Im weiteren Entwicklungsverlauf werden die Zellen des Augenmuskelkerns bis Em 20 derart von Migrationszellen umgeben, daß nur noch mit Hilfe der Fasern eine Lokalisation möglich ist. Em 23 zeichnet sich durch eine klare Zunahme der differenzierten Neurone dieses

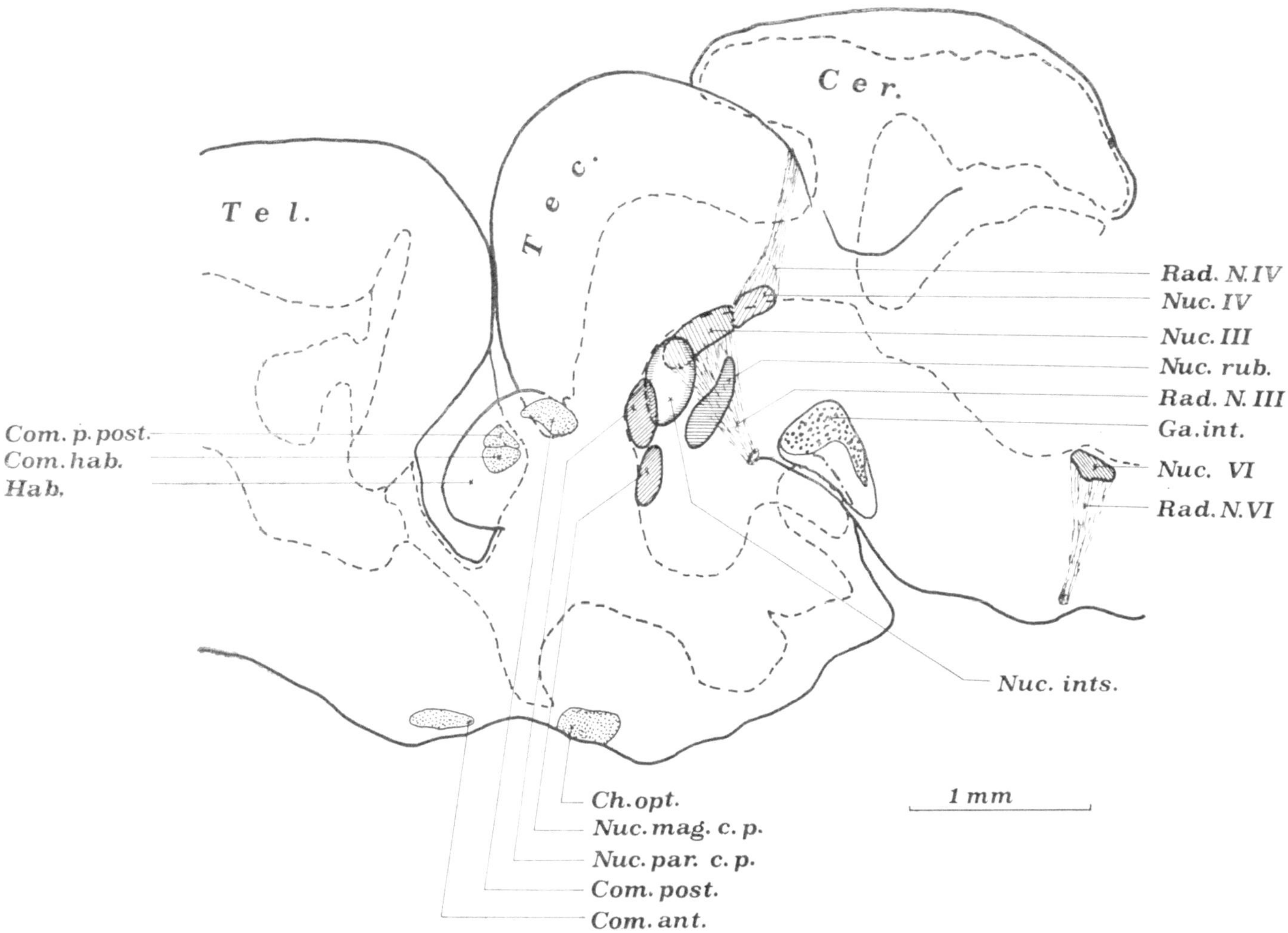

Abb. 3. Lateralprojektion von Mittelhirn und Zwischenhirn mit Augenmuskelkernen und Nuclei im mesencephalen Tegmentum. Das Ganglion interpedunculare ist mit seinem zellulären und faserhaltigen Anteil dargestellt.

Kerns aus. Auch sind die Fasern deutlich myelinisiert. In Em 28 wird eine Faserverbindung mit dem periventriculären Fasersystem des Hypothalamus evident. Kaudalwärts treten erste Faserkreuzungen des Nervus oculomotorius auf. Die Perikaryen erweisen sich durchweg multipolar. In Em 35 wird die Schichtung des Nucleus oculomotorius, welche über das Schlüpfstadium hinaus erhalten bleibt, offensichtlich; d. h. es kann das Kernareal in die Zonen 3, 4 und 5 unterteilt werden. Diese Gliederung wird später fast völlig verwischt.

Die Wurzel des Nervus oculomotorius liegt kaudal des Kernes, d. h. dorsal über der Wurzel liegt bereits der Nucleus trochlearis. Über der Commissura tegmentalis findet sich eine ventralwärts gerichtete Ausstülpung des Ventrikels. In diesem Bereich sind die Ependymzellen derart umgestaltet, daß sie sich bezüglich Form und Anordnung mit denen des Organon subcommissurale vergleichen lassen; diese Ähnlichkeit bleibt auch im erwachsenen Zustand erhalten. (Abb. 4E)

B) Nucleus trochlearis

Gleichsam als kaudale Verlängerung des Nucleus oculomotorius kann der Trochleariskern angesehen werden. Durch kräftige Faserzüge, die die Perikaryen beider Areale umgeben, werden die beiden Kerne eng miteinander verbunden. Die Zellen sind multipolar und entsprechen in Größe und Gestalt denjenigen des Nucleus oculomotorius. Ventrikelwärts legt sich eine Reihe kleinerer Zellen an, deren Fortsätze gegen das Kerninnere verlaufen. Axone, die die Wurzel des Nervus trochlearis bilden, ziehen von den Zellkörpern ventro-lateralwärts bis in den Fasciculus longitudinalis medialis, biegen dort rechtwinklig um und verlaufen in der Zone 6 nach dorsal. Die gut myelinisierten Fasern verbleiben, nach kaudal verlaufend, im Tegmentum mesencephali bis zum Isthmusgebiet, d. h. bis zum kaudalen Ende der tectalen Schichten 3 bis 6, wo sie als periventrikuläre Züge

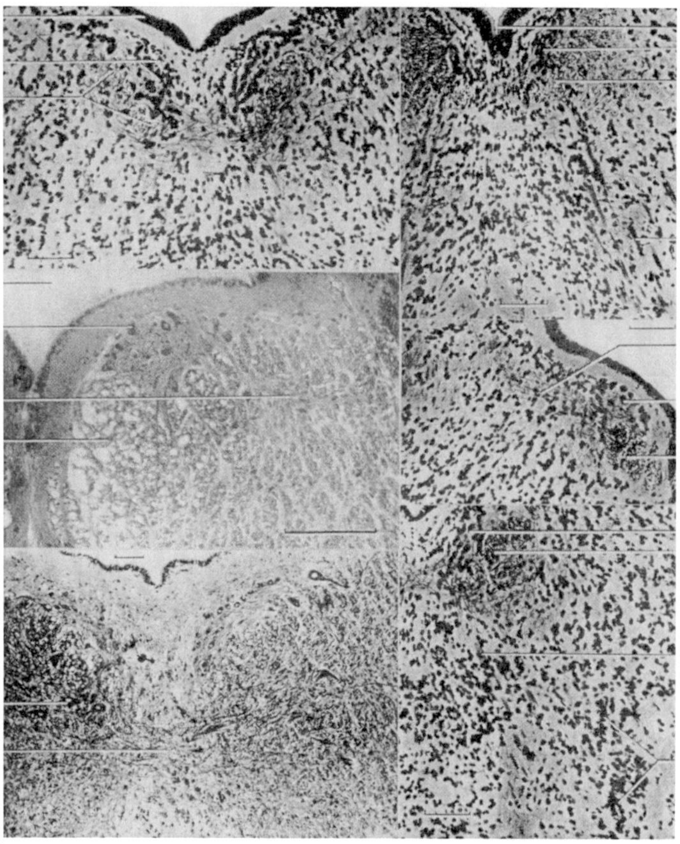

Abb. 4. Querschnitte durch das mesencephale Tegmentum. A: rostraler Nucleus oculomotorius mit rostralem F. l. m.; B: Wurzel des oculomotorius durchdringt den F. l. m., Org. s. = Zellen die morphologisch denjenigen des Organon subcommissurale entsprechen (vgl. Com. tegmentalis); C und D: Nucleus trochlearis dessen Fasern diejenigen des F. l. m. durchdringen und in Zone 6 nach dorsal verlaufen; E: Commissura tegmentalis und Nucleus interstitialis des F. l. m. setzen in gleiche Ebene ein; F: Nucleus ruber im Bereich der Oculomotoriuswurzel. (Maßeinheit = 100 μ)

zur Trochleariskreuzung ziehen und zwischen Cerebellum und Kaudalpol des Tectum opticum als Nervus trochlearis austreten.

Entsprechend der geringeren Mächtigkeit des Nucleus IV im Vergleich zum Nucleus III setzt auch die embryonale Entwicklung später ein. Erst mit Em 16 lassen sich eindeutige Bildungen erkennen; wobei bereits im Zusammenhang mit der Differenzierung des Nucleus III kaudal Trochleariszellen gebildet werden können, die lediglich nicht als solche erkennbar sind. Fasern werden bis Em 20 kräftig myelinisiert, kreuzen sich am dorso-kaudalen Tectum opticum und treten als Nervus trochlearis aus, wo sie, bereits von wenigen Gliazellen begleitet, zur Augenhöhle führen. Distal nimmt die Zahl der Gliazellen zu, sie umgeben die Fasern gleichsam mantelförmig. Die Axone können bis zu den einzelnen Fasern des Musculus obliquus superior beobachtet werden. Bis zum Schlüpfstadium findet eine Zunahme der Neurone und Fasern statt. Auch werden die Axone distal von Gliazellen begleitet, so daß diese den Fasern zwischengelagert sind. Bis zum ausdifferenzierten Zustand findet eine stärkere Trennung von Nucleus oculomotorius und Nucleus trochlearis statt. Der letztere wird, bedingt durch die Ausdehnung des Fasciculus longitudinalis medialis, nach dorso-lateral verdrängt und gelockert, derart daß teilweise Neurone im F. l. m. liegen. (Abb. 4C, D)

C) *Nucleus abducens*

Obwohl der Abducenskern bereits im rhombencephalen Bereich liegt, soll er als Augenmuskelkern hier in vorliegendem Zusammenhang beschrieben werden. Er ist kleiner als die zwei rostraleren Nuclei trochlearis und oculomotorius. Auch sind die an sich gleichgestalteten Zellen etwas kleiner. Der Kern liegt lateral des Fasciculus longitudinalis medialis, teilweise sogar eingestreut im Stratum griseum periventriculare. Obwohl die Augenmuskelkerne verschieden ausgebildet sind und rostrocaudal gesehen auseinanderliegen und besondere Arten der Nervenwurzeln aufweisen, können sie alle derselben Schicht zugeordnet werden. Die Wurzel des Nervus III besteht aus einzelnen dichtliegenden Faserbündeln; die Wurzel des Nervus VI aus ebensolchen, die jedoch in größeren Abständen zur ventralen Peripherie ziehen und dort eine breite Wurzelzone bilden; wo auch die Gliazellen faserbegleitend auftreten.

Ontogenetisch lassen sich erste Neurone des Nucleus abducens verhältnismäßig spät nachweisen; in Em 16 sind eingestreute, wahrscheinlich noch migrierende Zellen im Fasciculus longitudinalis medialis ersichtlich, deren Axone bereits einen sehr schmächtigen Nervus abducens bilden. Mit Em 26 stellt sich eine klare Separation zwischen periventrikulärem Fasersystem und dem anschließenden Grau ein. Die Neurone des Nucleus abducens liegen vorwiegend innerhalb des F. l. m., sind jedoch nahe der endgültigen Zahl vorhanden. Bis zum Schlüpfstadium ist die Faserzunahme erheblich; auch der Grad der Myelinisation, insbesondere der distalen Partien. Wie schon bei der Oculomotoriuswurzel kreuzen sich vereinzelte Fasern der Abducenswurzel, jedoch in wesentlich geringerer Zahl. Im differenzierten Zustand sind die einzelnen Neurone, bedingt durch Faserlockerungen, derart verstreut, daß in einem Querschnitt nur zwei bis drei sichtbar sind. Entsprechend ist die Wurzel beim adulten Tier um ein Mehrfaches größer als beim Schlüpfling.

D) *Nucleus interstitialis*

Der Nucleus interstitialis des Fasciculus longitudinalis medialis, auch als Nucleus motorius tegmenti bezeichnet, liegt ventrikelnahe im rostralen Tegmentum. Von ihm ziehen die stark myelinsierten kräftigen Fasern paramedian als Fasciculus longitudinalis medialis rostralwärts durch die Medulla oblongata ins Rückenmark (Abb. 5). Die sehr großen multipolaren Zellen, wie sie bereits Beccari (1923) anhand eines Reptilienhirns beschrieben hat, weisen Verbindungen mit dem periventrikulären Weiß des Tectums (Zone 2 und 4), jedoch auch mit der zentralen Zone 6 auf. Ventralwärts bestehen Faserkontakte zu paraventriculären Strukturen. Das kräftige Bündel des Fasciculus retroflexus endet mit einem Teil seiner Fasern im Nucleus interstitialis des F. l. m.; ein weiterer Anteil durchdringt das Kerngebiet und zieht kaudalwärts zum Nucleus interpeduncularis. Dendriten und deren Verästelungen strahlen radial zur Peripherie bis zum Nucleus lentiformis praetectalis. Im kaudalen Kerngebiet treten vereinzelte Dekussationsfasern auf, die innerhalb der Commissura tegmentalis verlaufen. Faserkontakte bestehen mit dem Rostralpol des Nucleus ruber.

Die embryonale Entwicklung des Nucleus interstitialis des Fasciculus longitudinalis medialis, setzt morphologisch feststellbar mit Em 9 ein. Die etwas größeren Zellen bilden eine kompakte periventrikuläre Lage. Mit Em 11 setzt die Differenzierung ein, die typische Form beginnt sich abzuzeichnen. Die Fasern des Fasciculus retroflexus reichen bereits bis in die Kernregion. Mit Em 13 wird ein enormes Wachstum der Perikaryen evident; die Zellkörper sind im Vergleich zu den Umliegenden bereits auf die dreifache Größe angewachsen. Die den Fasciculus longitudinalis medialis bildenden Fasern sind in Em 14 sehr deutlich, sie reichen kaudalwärts bereits über das mesencephale Tegmentum hinaus. In den Stadien bis Em 23 weisen fast alle Zellkerne eine annähernd ideale Kugelform auf; d. h. in Em 18 und 20 sind Hauptmigrationsphasen. Das Gebiet des Nucleus interstitialis ist reichlich mit Migrationszellen durchsetzt. Die zum Kerngebiet gehörenden Zellen sind lediglich an den erheblich größeren, eben-

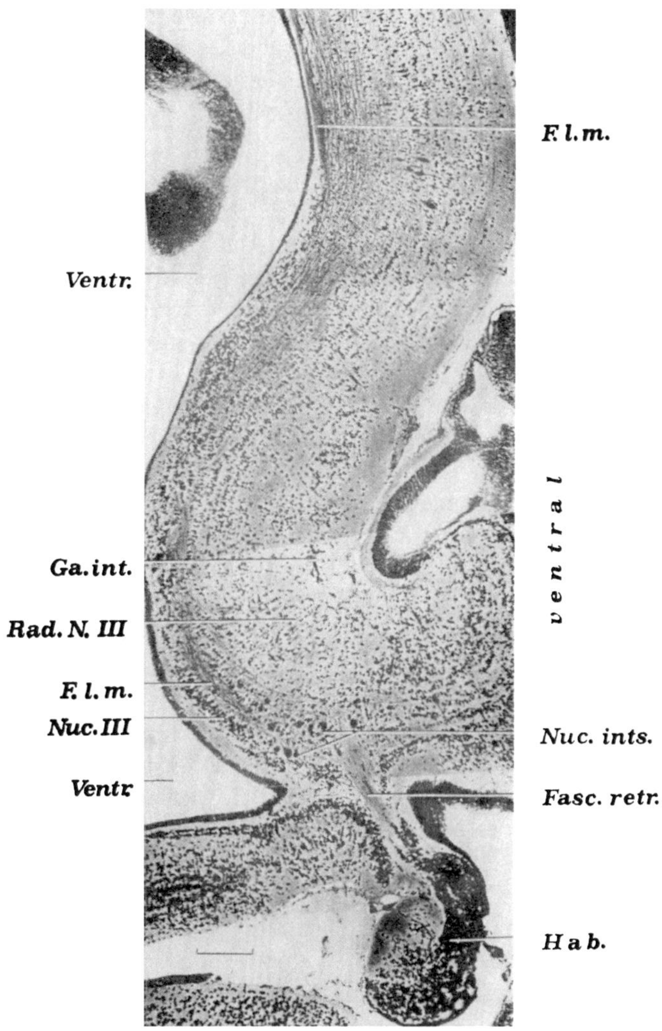

Abb. 5. Paramedianschnitt durch das mesencephale Tegmentum. Darstellung von Nucleus interstitialis, Fasciculus longitudinalis medialis und Fasciculus retroflexus und deren Zusammenhänge. (Maßeinheit = 200 μ)

falls kugeligen Zellkernen auszumachen. Die Faserbündel des Fasciculus retroflexus werden mit Em 23 auch von der Ventralseite mit Zellen des Nucleus interstitialis umlagert. Kaudal des Kerns zieht der Fasciculus retroflexus noch mit knapp halber Stärke weiter und zweigt sich im Ganglion interpedunculare auf. Struktur und Aufbau des Kernes haben mit Em 25 nahezu den Stand des Schlüpflings erreicht. Bis Em 35 bilden die ersten Wurzelfasern des Nervus oculomotorius gleichsam den caudalen Abschluß des Nucleus interstitialis, dessen Neurone mit Em 36 in den Bereich der Oculomotoriuswurzel vordringen und bis zum Neonatus lediglich innerhalb der rostralen Wurzelzone auftreten. Eine kaudale Ausdehnung erfährt der Kern im postnatalen Zustand. Auch die bilaterale Ausweitung ist evident. In der späten Wachstumsphase erfährt der lockere Nucleus interstitialis eine erhebliche Auflockerung der Zelldichte; wobei die dorso-ventrale Ausdehnung des Kernes postnatal nur unwesentlich verändert wird. Auffällig ist die Perikaryengröße des ausdifferenzierten Kerns, die im Vergleich zu derjenigen des Schlüpfstadiums auf das ca. fünf- bis achtfache Volumen angestiegen ist. (Abb. 5)

E) Nucleus ruber

Der Nucleus ruber liegt kaudal des Nucleus interstitialis des Fasciculus longitudinalis medialis, nach ventrolateral verschoben. Mit dem kaudalen Teil liegt er lateral der Oculomotoriuswurzel, deren Fasern den Nucleus ruber teilweise durchziehen und vereinzelt in Kontakt treten. Die Lagebeschreibung von De Lange (1913), der Nucleus ruber befinde sich lateral neben der Wurzel des Nervus oculomotorius, deckt sich mit den vorliegenden Befunden, obgleich der größere Teil die Wurzel rostral überflügelt.

Die polygonalen Zellen des Nucleus ruber sind mittelgroß, die Zellen weisen im Mittel einen Durchmesser von 20—30 µm auf, wogegen diejenigen des Nucleus oculomotorius durchschnittlich 40 µm messen. Beccari (1933) beschreibt die Zelle der N. ruber bei Teleostei als ovale Zellen, die der Größe eines Neurons des Nucleus oculomotorius entsprechen. Im ventralen Anteil der Commissura tegmentalis kreuzen sich, die vom Nucleus ruber medianwärts wegziehenden Fasern, die ebenfalls Nucleus III Axone orthogonal kreuzen. Weniger dicht verlaufen nervöse Verbindungen nach dorso-kaudalcerebellumwärts.

In Em 16 sind die meisten Zellen des Nucleus ruber bereits aus der Proliferationszone migriert. Sie liegen jedoch noch derart dicht ventral des Nucleus oculomotorius, daß sie als kaum differenzierte Zellen, nicht eindeutig als Zellen des Nucleus ruber auszumachen sind. Zu einem Kern gleichgestalteter Neurone formieren sie sich in Em 23, jedoch mit noch allseitig fließendem Übergang in angrenzende Zonen. Der typisch polygonale Zellkörper entwickelt sich sporadisch bis Em 26 und ist bereits in Em 30 im ganzen Nucleus evident. Die medianwärts ziehenden Efferenzen werden erst mit Em 39 evident, obwohl die Myelinisation auch beim Neonatus noch schwach ist und postnatal wesentlich verstärkt wird. Der im Schlüpfling noch dicht erscheinende Kern wird bis zur endgültigen Ausbildung stark aufgelockert, auch werden die Perikaryen etwas größer. (Abb. 4F)

F) Ganglion interpedunculare

Das Ganglion interpedunculare kann in einen dorsalen Nucleus und ein ventrales plexiformes Fasergebiet gegliedert werden; wobei sich der zelluläre Teil und die Fasern derart durchdringen, daß eine Trennung wenig sinnvoll ist. Das Ganglion liegt ventro-median im Tegmentum und beginnt rostral zwischen den beiden Wurzeln des Nervus oculomotorius und reicht kaudalwärts bis zum distalen Hypophysenpol.

Das Ganglion gliedert sich kaudalwärts deutlich in einen zellulären und einen faserführenden Teil, der kaudal über dreiviertel des Ganglion ausmacht. Die kugeligen Zellen liegen dicht, sie erlauben eine exakte Arealabgrenzung.

Die mannigfaltigen Faserverbindungen lassen sich nur schwer feststellen, da der Myelinisationsgrad klein ist und die Fasern zahlreich sind. Sehr eindeutig, bereits früher erwähnt, ist die Verbindung mit der Habenula durch den Fasciculus retroflexus, der sich kaudal des Nucleus interstitialis des F. l. m. aufsplittert und den Rostralpol des Ganglion interpedunculare auf breiter Basis erreicht.

Mit Em 13 migrieren vereinzelt Zellen in das Gangliongebiet. Ventral sind erste plexiforme Fasermassen erkenntlich. Mit Em 16 wandern bereits weitgehend differenzierte Zellen des Nucleus interpenduncularis paramedian ventralwärts und gruppieren sich in dichter Formation. Mit Em 23 wird evident, daß der zelluläre Anteil dem anderen weit vorangeht. Dieser Vorsprung beginnt sich ab Em 28 auszugleichen, in einem Stadium also, in dem allgemein verstärktes Faserwachstum einsetzt. Wie bereits in anderen Nuclei des mesencephalen Tegmentums festgestellt, erfährt auch der zelluläre Anteil des Ganglion interpedunculare in der postnatalen Entwicklungsphase eine starke faserbedingte Lockerung neben einer durch commissurale Fasern erwirkten horizontalen Schichtung.

G) Kerne der Commissura posterior

Dorso-kaudal der Commissura habenularis schließt die Commissura posterior an. Sie liegt den langgestreckten Zellen des Organon subcommissurale als Faserzone auf und wird am kaudalen Ende von dorsal gelegenen Subcommissuralzellen, dem Recessus pinealis, umfaßt. Die Commissurenfasern nehmen an Myelinisation zu, ziehen in kaudoventraler Richtung lateralwärts, und durchdringen dabei den Nucleus parvocellularis commissurae posterioris. Dieser Kern grenzt dorsal an den Fasciculus retroflexus und liegt über der Pars plicata des Nucleus lentiformis praetectalis. Axone dieser mittelgroßen polygonalen Zellen biegen rechtwinklig um und verlaufen tangential dorsal oder ventralwärts. Die Dendriten spalten sich mannigfaltig auf. Fasern der Commissura posterior die ventrikelparallel ventral-kaudal verlaufen, ziehen zum Nucleus magnocellularis commissurae posterioris. Dieser tritt mit dem Verschwinden der Pars plicata gleichsam an dessen Stelle kaudalwärts auf. Die großen polygonalen Zellen formieren sich ventral des Fasciculus retroflexus zu einem lockeren Kern. Im Vergleich mit den Zellen des Nucleus interstitialis des Fasciculus longitudinalis medialis besitzen die Neurone des Nucleus magnocellularis in etwa Zweidrittel der Größe von Ersteren. Die Dendriten verlaufen vorwiegend radial, die Axone tangential. Beccari (1923, 1943) hat die Kerne bei *Lacerta muralis* beschrieben, wobei sich bezüglich Lage und Zelldifferenzierung gewisse Ähnlichkeiten ergaben.

Ontogenetisch geht die Commissura posterior den beiden Kernen weit voran. In Em 11 sind bereits myelinisierte Fasern, dorsal der Commissura habenularis, ausgebildet. In Em 20 lassen sich Zellen des Nucleus magnocellularis ausmachen, die in Em 23 deutlich differenziert sind und in Em 26 die Größe von denjenigen des Schlüpflings aufweisen. Die Neurone des Nucleus parvocellularis sind ihrer mittleren Größe wegen, erst erkennbar, wenn sie ausdifferenziert sind, d. h. die Migration findet in den Schubphasen von Em 18—20 statt. Die typische Gestalt der Zellen wird erst mit Em 26 evident, wobei die dem Neonaten vergleichbare Größe mit Em 33 erreicht ist. Postembryonal werden die Kernareale erwartungsgemäß aufgelockert, die Perikaryen nehmen an Größe proportional denjenigen anderer tegmentaler Kerne zu. Auch werden die Fasern stärker myelinisiert.

Abkürzungen

Cer	Cerebellum
Ch. opt.	Chiasma opticum
Com. ant	Commissura anterior
Com. hab.	Commissura habenularis
Com. post.	Commissura posterior
Com. p. post.	Commissura pallii posterior
Com. teg.	Commissura tegmentalis
diff. Ze.	differenzierende Zellen
Ep.	Ependym
Fasc. retr.	Fasciculus retroflexus
F. l. m.	Fasciculus longitudinalis medialis
Ga. int.	Ganglion interpedunculare
Ga. Ze.	Ganglienzellen
Gl. ze.	Gliazellen
Hab.	Ganglion habenulare
L.	Linse
Mig. ze.	Migrationszellen
Mit. ze.	Migrationszellen
Mit. ze.	Mitosezellen
Nuc. ints.	Nucleus interstitialis des Fasciculus longitudinalis medialis
Nuc. mag. c. p.	Nucleus magnocellularis commissurae posterioris
Nuc. III	Nucleus oculomotorius
Nuc. IV	Nucleus trochlearis
Nuc. VI	Nucleus abducens
Nuc. par. c. p.	Nucleus parvocellularis commissurae posterioris
Nuc. rub.	Nucleus ruber
Pig. ep.	Pigmentepithel
Rad. N. III	Radix nervus oculomotorius
Rad. N. IV	Radix nervus trochlearis
Rad. N. VI	Radix nervus abducens
Ret.	Retina
Se. ze.	Sehzellenschicht
S. grau. in.	innere Körnerschicht
S. plex. ex.	äußere plexiforme Schicht
S. plex. in.	innere plexiforme Schicht
Sts.	Stäbchensaum
Tec	Tectum opticum
Tel	Telencephalon
Tr. opt.	Tractus opticus
Ventr.	Ventrikel

Literatur

Alluchon-Gerard, M. J. et J. Mellinger: Mise en évidence de plusieurs phases distinctes dans la croissance embryonnaire de la rousette (Scullium canicula CUV.) Ann d'Embr. et. Morph. **4**, 19—35, (1971)

Ariens Kappers, C. U., G. C. Huber and E. C. Crosby: The comparative anatomy of the nervous system of vertebrates including man (Hafner, New York 1960)

Ariens Kappers, C. U.: Teleostean and selachian brain. Jour. comp. Neur. **16**, 1—112 (1906)

Beccari, N.: Il centro tegmentale o interstiziale et altre formazioni poco note nel mesencefalo e nel diencefalo di un rettile. Arch. ital. Anat. Embriol. **20**, 560—619 (1923)

Beccari, N.: Existe-t-il un noyan rouge chez les poissons téleosteens? C. R. Ass. Anat., Réun. **28**, p. 51 (Lisbonne) (1933)

Beccari, N.: Neurologia comparata anatomo-funzionale dei vertebrati, compreso l'uomo (Sansoni, Firenze 1943)

Bergquist, H.: Ontogenesis of diencephalic nuclei in vertebrates. Lunds Universitets Arsskrift **50**, 1—33 (1954)

Burckhardt, R.: Das Zentralnervensystem der Selachier I. Teil: Acta Leopoldino-Carolinae nat. curios 73 (1907) II. Teil daselbst Bd. 94 (1911)

Ebbesson, S. O. E. and S. J. Ramsey: The optic tracts of two species of Sharks (Galeocerdo cuvier and Ginglymostoma cirratum). Brain Res. **8**, 36—53 (1968)

Ebbesson, S. O. E.: New insights into the organisation of the shark brain. Comp. Biochem. Physiol. 42A, 121—129 (1972)

Ebbesson, S. O. E. and C. B. G. Campbell: On the organisation of Cerebellar Efferent Pathways in the Nurse Shark (Ginglymostoma cirratum) J. comp. Neur., 152, 233—254 (1973—74)

Farner, H.-P.: Untersuchungen zur Embryonalentwicklung des Gehirns von Scyliorhinus canicula (L.): I. Bildung der Hirngestalt, Migrationsmodi und -phasen, Bau des Zwischenhirns. — II. Das Tectum opticum und dessen Stratifikation.

Franz, V.: Die Akkomodation des Selachierauges und seine Abblendungsapparate, nebst Befunden an der Retina. Zool. Jb. allg. Zool. Physiol. **49**, 323—462 (1931)

Haller, B.: Vom Bau des Wirbeltier-Gehirns. I. Salmo und Scyllium. Morph. Jb. **26**, 345—641 (1898)

Kuhlenbeck, H.: The central nervous system of vertebrates, Vol **4** (Karger, Basel, New York 1975)

Kupfer, K. von: Die Morphogenie des Zentralnervensystems; in Hertwig Hb. der vergl. experiment. Entwicklungslehre der Wirbeltiere Vol. 2, Teil 3, 1—119 (Fischer, Jena.1906)

Kusunoki, T., Y. Tsuda and F. Takashima: The chemoarchitectonics of the shark brain. J. Hirnforsch. **14**, 13—26 (1973)

Lange, S. J. de: Das Zwischenhirn und das Mittelhirn der Reptilien. Folia neurobiol. **7**, 67—138 (1913)

Marinelli, W. und A. Strenger: Vergleichende Anatomie und Morphologie der Wirbeltiere, III. Lieferung, Superklasse Guathostomata, Chondrichthyes (Deuticke, Wien 1969)

Polyak, S.: The vertebrate visual system. University of Chicago Press. Chicago, Illinois (1957)

Ramon, P.: Estructura del encefalo del camaleon. Rev. trimest. micrograf. **1**, 46—82 (1896)

Rochon-Duvigneaud, A.: Les yeux et la vision des vetérbrés. Masson, Paris (1943)

Scammon, R. E.: The development of Squalus acanthias, in Normentafeln zur Entwicklungsgeschichte der Wirbeltiere Heft 12, Keibel, Jena 1911

Senn, D. G.: Über das optische System im Gehirn squamater Reptilien. Acta anat., Suppl. 52 = 1 ad 65 (1966)

Senn, D. G.: Bau und Ontogenese von Zwischen- und Mittelhirn bei Lacerta sicula (Rafinesque) Acta anat. Suppl. 55 = 1 ad 71: 1—150 (1968a)

Senn, D. G.: The stratification in the reptilian central nervous system. Acta anat. 75, 521—552 (1970)

Senn, D. G.: Notes on the amphibian and reptilian thalamus. Acta anat. **87**, 555—596 (1974)

Stell, W. K. and P. Witkovsky: Retinal Structure in the Smooth Dogfish, Mustelus canis: General description and light Microscopy of giant ganglion cells. J. comp. Neurol. **148**, 1—32 (1973)

Stell, W. K. and P. Witkovsky: Retinal Structure in the Smooth Dogfisch, Mustelus canis: Light Microscopy of Photoreceptor and horizontal cells. J. comp. Neurol. **148**, 33—46 (1973)

Witkovsky, P. and W. K. Stell: Retinal Structure in the Smooth Dogfish, Mustelus canis: Light Microscopy of bipolar Cells. J. comp. Neurol. **148**, 47—60 (1973)

Adresse des Autors:

Dr. H.-P. Farner
Zoologisches Institut der Universität Basel
4051 Basel, Rheinsprung 9
Switzerland

J. Hirnforsch. 19 (1978) 415—421

From the Paul-Flechsig-Institute of Brain Research, Leipzig

Vesicle size in basic classes of synapses in the rat's lateral geniculate nucleus[1]

By Dmitrij N. LENKOV[2], Ernst WINKELMANN, Kurt BRAUER and Gert BRÜCKNER

With 3 figures

(Received September 15, 1977)

Summary: Five basic classes of synaptic terminals within the dorsal lateral geniculate nucleus of the albino rat were identified on electron-micrographs according to conventional qualitative criteria. Two statistical parameters characterizing adequately the diameter of synaptic vesicles located in the terminals of each class were determined. Significant differences in synaptic vesicle size were stated at least in three classes of synaptic terminals in the dorsal lateral geniculate nucleus.

Резюме: На электронномикроскопических фотографиях тонких срезов дорзального латерального коленчатого ядра белой крысы с помощью принятых качественных критериев идентифицировали пять основных классов синаптических терминалей. Определяли также два статистических параметра, адекватно характеризующих диаметр синаптических везукул, расположенных в терминалях каждого класса. Установлены достовреные различия размера синаптических везикул по крайней мере в трёх классах синаптических терминалей в дорзальном латеральном коленчатом ядре.

Zusammenfassung: Fünf Grundtypen synaptischer Terminalen wurden im dorsalen Corpus geniculatum laterale der Albinoratte auf Grund konventioneller qualitativer Kriterien elektronenmikroskopisch identifiziert. Zur Bestimmung des Durchmessers der synaptischen Vesikel jedes Terminaltyps wurden zwei statistische Parameter ausgewertet. Nach der Größe der synaptischen Vesikel wurden signifikante Unterschiede mindestens zwischen drei Typen synaptischer Terminalen des dorsalen Corpus geniculatum laterale festgestellt.

Current morphological investigations of the dorsal lateral geniculate nucleus (dLGN) make it possible to identify basic different classes of synapses basing on their ultrastructural peculiarities. Following citeria are widely used for classification, i.e. the shape of synaptic vesicles, the size of terminals, the presence or absence of mitochondria, the specific localization of terminals on the postsynaptic membrane (on the soma or on the dendritic branches and spines) and the type of synaptic contact, symmetrical or asymmetrical after GRAY-COLONNIER's classification.

On the base of all these criteria it is possible to differentiate within the albino rat dLGN at least 4—5 several classes of synaptic terminals (MONTERO and GUILLERY, 1968; GUILLERY, 1969; BRÜCKNER, BAER and BIESOLD, 1972; LIEBERMAN and WEBSTER, 1972, 1974; BRAUER and WINKELMANN, 1974; BRAUER, WINKELMANN, MARX and DAVID, 1974; BOGOLEPOV, 1975), including a particular class of terminals of probable dendritic origin which were called ID-terminals (FAMIGLIETTI and PETERS, 1972; LIEBERMAN and WEBSTER, 1972).

However, in spite of the relatively simple intrinsic organization of the dLGN in rodents, certain identification of individual terminals and, moreover, the determination of their origin from different intra- and extrageniculate sources (retinal, cortical, brain stem, other thalamic nuclei etc.) brings about some difficulties (BRÜCKNER et al., 1972; LIEBERMAN and WEBSTER, 1974).

Recently, among additional criteria for identifying synaptic terminals of different origin, the diameter of synaptic vesicles in various structures of the central and peripheral nervous systems in vertebrates and invertebrates has attracted close attention (e.g. EVANS, 1966; KAWANA, AKERT and BRUPPACHER, 1971; FROESCH and MARTIN, 1972; LENKOV, BRAGINA and MALOSHEVSKY, 1974, 1975).

It is known that synaptic vesicle size depends on a number of factors. Among such are the level of the functional activity in the course of the brain tissue

1 This investigation was sponsored by a grant from the Ministry of Science and Technology of the G.D.R.

2 Dr. D. N. LENKOV has been working at Leipzig University as IBRO-Stipendiate. Present address: Physiological Institute, Leningrad State University, USSR.

fixation (Korneliussen, 1972; McKinlay and Usherwood, 1973; Zimmermann and Whittaker, 1974; Maloshevsky, Lenkov and Bragina, 1975), the stage of maturation of synapses (del Cerro and Snyder, 1969) and degeneration processes (Cuenod, Sandri and Akert, 1970; Kawana et al., 1971). Yet, abundant data show that the synaptic terminals derived from the same neuronal population in adult and healthy animals are characterized in the resting state by relative stable parameters of synaptic vesicle size (Lenkov, 1972; Froesch, 1973; Lenkov et al., 1974). Correspondingly, significant differences were observed in the synaptic vesicle diameter in different classes of synaptic terminals in some parts of the nervous system in various animal species (Evans, 1966; Kawana et al., 1971; Froesch and Martin, 1972; Lenkov et al., 1975).

Therefore, the present study was mainly aimed at the determination of the diameters of synaptic vesicles belonging to different classes of synaptic terminals in the albino rat dLGN by means of the technique introduced recently (Lenkov et al., 1974; Maloshevsky et al., 1975).

Material and Methods

Electron microscopy

Brains of albino rats were fixed by perfusion with a mixture of 5% glutaraldehyde and 4% paraformaldehyde in 0,1 M sodium phosphate buffer, pH 7,2—7,4. The lateral geniculate nucleus was removed, cut in little blocks and subjected to postfixation with the perfusion mixture for 1,5 hour and with 1% OsO_4 for one hour. After dehydration the tissue blocks were embedded in Micropal. The thin sections were stained with uranyl acetate and lead citrate. Photographs were taken with the electron microscope SEM 3—1 VEB Fernsehelektronik Berlin.

Quantitative procedures

Vesicle diameter in electronmicroscopical plates got at 15.000-fold magnification were measured through the reader-device Dokumator DL4 (Zeiss, Jena) with an additional 21-fold magnifikation. Since the true synaptic vesicle size is characterized by the population of largest vesicular profiles in a section (Lenkov, 1972; Froesch, 1973), synaptic vesicle diameter in each terminal (a half-sum of two mutual-perpendicular diameters, large and small) was measured only at 15—25 synaptic vesicles. The number of vesicles measured depended on the total number of those in the section. Such were included in the frequency histograms of synaptic vesicle size, which were plotted for each individual terminal and were further analyzed.

The synaptic vesicle size at the terminal was characterized also by means of two statistical parameters, i.e. the so-called submaximum decile, or percentile (p-parameter) and the maximum significant diameter (D_{max}). The profile of synaptic vesicle with largest diameter within the given terminal was taken as D_{max} (Lenkov et al., 1974). Parameter p is one of the order statistics (van der Warden, 1960) which was determined along the following rule. — If there is a given variation succession $x_1 \geqq x_2 \cdots \geqq x_n$, with $n = 10k$, the member x_{k+1} is taken as p. When $n = 10k + s(1 \leqq s \leqq 9)$, linear interpolation was used to determine the value of p from x_{k+1} and x_{k+2} (for detailes see Maloshevsky et al., 1975).

Results

Classification of terminals

Within the dLGN presynaptic profiles containing electrontransparent synaptic vesicles were found predominant. Terminals loaded with synaptic vesicles of that kind were investigated in details. The axons and varicosities loaded with dense core vesicles were observed rarely and no cases of a synaptic contact formed by such elements in the dLGN's neuropil was found. As follows from the above-mentioned qualitative criteria, five types of synaptic terminals can be distinguished in the dLGN.

Type 1. Axon terminals of the highest frequency are distributed almost homogeneously over the entire volume of the dLGN and they have the smallest diameter ranging from 0,3—0,8 μm. These terminals do not contain mitochondria and form synapses at distal portions of thin dendritic branches of the dLGN's neurones predominantly (Fig. 1a).

Type 2a. (so-called R-boutons). The largest axon terminals within the dLGN with a diameter ranging 2—5 μm contain usually several mitochondria (Fig. 1b). These terminals form multiple non-synaptic contacts of a filamentous, desmosome-like type along the dendritic branches of the principal dLGN cells, i.e. geniculocortical relay neurones, but the same terminals form true synaptic contacts on some dendritic protrusions (grape-like appendages). As it has been shown by a number of investigators, that the terminals of type 2a are enveloped by membranes of glia cells forming a peculiar lamellar capsule. These axon terminals are presynaptic to so-called ID-terminals, containing disc-like vesicles.

Type 2b. The axon terminals of the medium size (diameter 1—2 μm), with one or more mitochondria and with one active zone mostly connected to distal dendrites (Fig. Ic).

Type 3. (so-called F-boutons, IA-terminals). This class of axon profiles is very similar to the profiles of type 2b in some general morphologic features: they have the same size range about 1 μm and contain mitochondria. However, the terminals of type 3 are distributed on neuronal bodies and proximal parts of dendrites. The synaptic contacts correspond to the symmetrical type and the synaptic vesicles

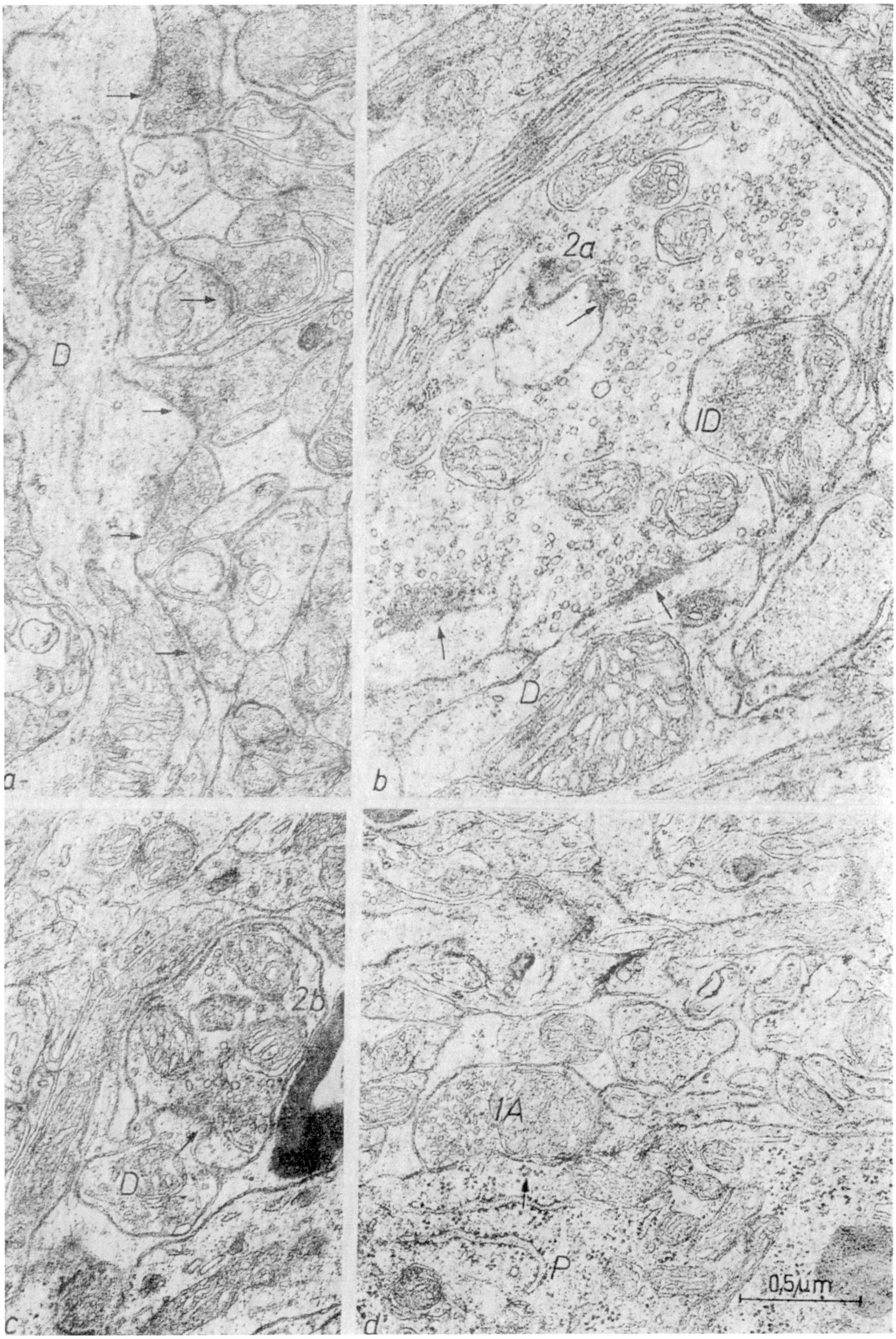

Fig. 1. Electron micrographs of the typical examples of the presynaptic terminals of five identified classes within the dorsal lateral geniculate nucleus in the albino rat. a, type 1 terminals; b, type 2a terminal and type 4 terminal (ID-terminal); c, type 2b terminal; d, type 3 terminal (IA-terminal). Abbreviations: D, dendrite; P, perikaryon. Arrows indicate the active zone of synaptic contacts. Magnification × 43000.

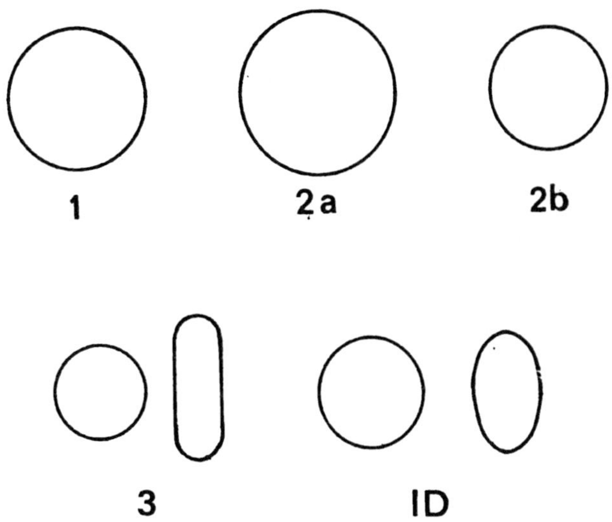

Fig. 2. Diagram of synaptic vesicles of different size and shape, which were observed in the corresponding classes of presynaptic endings: type I, type 2a, type 2b, type 3 (IA-terminals) and type 4 (ID-terminals). The last two categories of terminals both contain two different variants of vesicular profiles, round and flattened, but they are characterized by different ratio of large to small axes.

have a distinct cylindrical shape: a section of the same terminal represents both long rod-like vesicular outlines and small round ones (Fig. 1d and Fig. 2).

Type 4. (so-called P-boutons, ID-terminals). These terminals, probably dendritic, contain synaptic vesicles of disc-like shape. The latter is recognized by the co-existence of two variations of vesicular profiles within the same terminal, round and oval, with similar long axis (Fig. 1b and Fig. 2).

Size of vesicles

Quantitative estimations performed on 60 identified presynaptic profiles indicate, that among the five classes of synaptic terminals in the dLGN terminals of type 2a (R-boutons) and terminals of type 3 (IA-terminals) represent two extreme variants according to the synaptic vesicle size (Fig. 3). The values of statistical parameters were estimated for terminals of type 2a as follows: $p = 54-57$ nm and $D_{max} = 60-64$ nm, and for terminals of type 3: $p = 31$ to 36 nm and $D_{max} = 39-44$ nm. The examples of individual frequency histograms and of values of D_{max} and p are presented on Fig. 3b and 3d, correspondingly. The difference between the two above classes of the presynaptic endings exceeded 60 to 75 per cent for both parameters, that can be considered as highly demonstrative.

The size of disc-like synaptic vesicles within the ID-terminals which in relation to the terminals of type 2a are postsynaptic structures is much smaller than the size of round synaptic vesicles loading the terminals of type 2a. It is evident from a comparison of distribution and modal intervals in frequency histograms (Fig. 3b and c). Statistical parameters characterizing the size of disc-like vesicles were determined as follows: $p = 39-43$ nm and $D_{max} = 49-51$ nm. These are 35–40 per cent less than corresponding values of p-parameter and D_{max} obtained for the terminals of type 2a, but, on the other hand, these are 20–25 per cent higher than the values of both parameters in the terminals of type 3 (cf. also the frequency histograms on Fig. 3c and d).

A less clear picture was revealed during the analysis of the ultrastructure of the type 1 and the type 2b terminals. It was connected with the difficulty of exact distinguishing these two classes of presynaptic endings using the generally accepted criteria.

The values of parameters characterizing the size of synaptic vesicles in the type 1 terminals were determined $p = 43-49$ nm and $D_{max} = 47-57$ nm. Corresponding values for terminals of type 2b were $p = 41-49$ nm and $D_{max} = 49-54$ nm.

Discussion

The classification of synapses in the dLGN given in this paper is in accordance with the previous findings (Guillery, 1969; Brückner et al., 1972; Famiglietti and Peters, 1972; Lieberman and Webster, 1972, 1974; Giolli and Pope, 1973; Brauer and Winkelmann, 1974; Brauer et al., 1974; Bogolepov, 1975).

Five distinct classes of presynaptic profiles could be differentiated: type 1, type 2a, type 2b, type 3 (IA-terminals) and type 4 (ID-terminals). There exist some indications concerning the origin of those presynaptic profiles. Extensive degeneration of type 1 axon terminals occuring after the destruction of visual cortex indicate that these terminals originate from cortico-geniculate fibres (Montero and Guillery, 1968; Giolli and Pope, 1973). It can be supposed that the great majority of type 2a and 2b terminals are of retinal origin. Lieberman and Webster (1972, 1974) do not subdivide the terminals containing round vesicles into two such subgroups and call them R-boutons. Brauer et al. (1974) considered 2a and 2b terminals as close subclasses. It was also found that the subclass 2a terminals are concentrated in two rostral thirds of the dLGN, however, they are extremely rare in the caudal part of the nucleus (Brauer and Winkelmann, 1977). On the other hand, type 2b terminals are distributed over

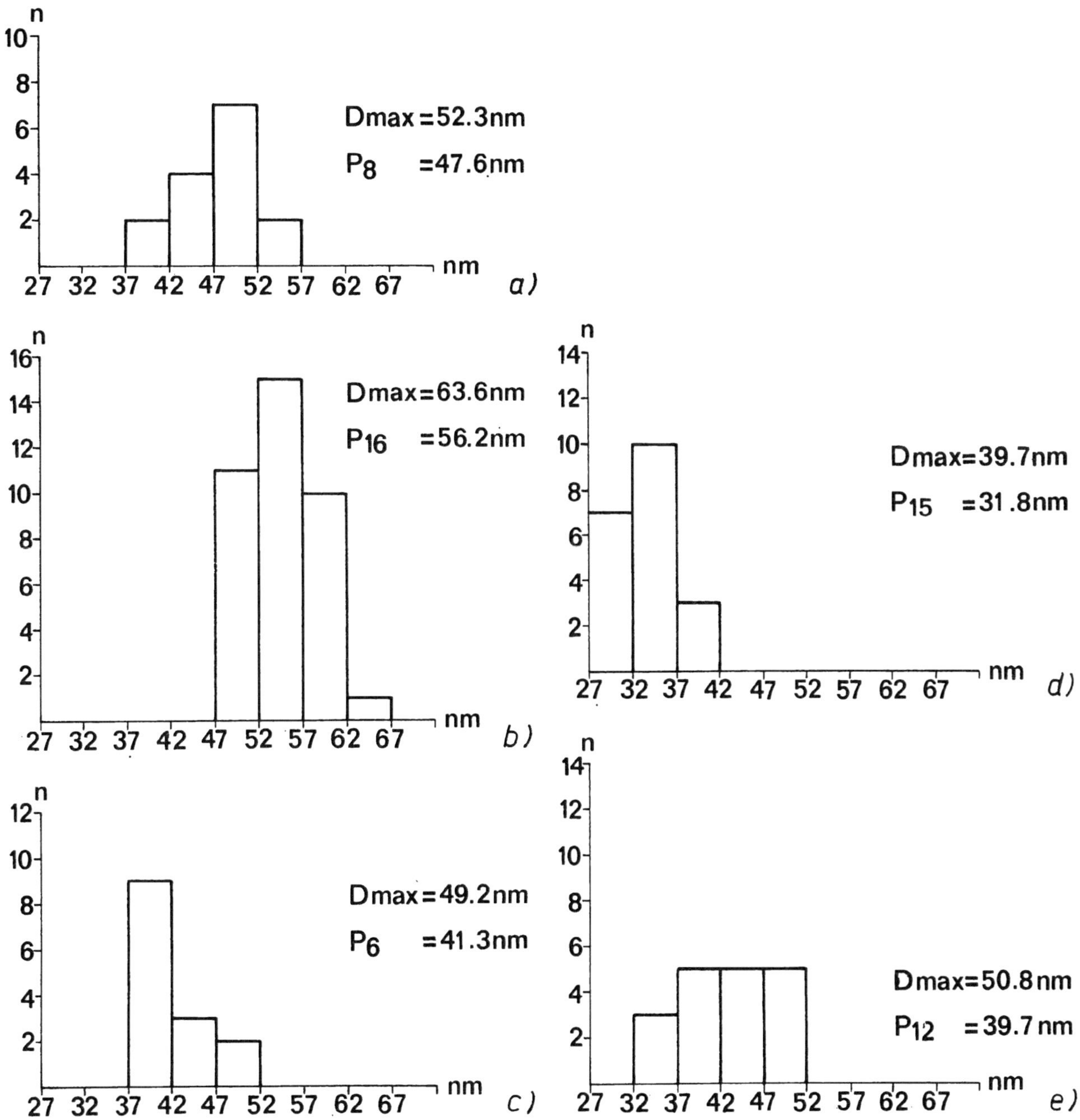

Fig. 3. The frequency histograms of synaptic vesicle diameter determined in largest vesicles within given terminal. a—e, the frequency histograms for individual presynaptic terminals of five identified classes: type 1, type 2a, type 2b, type 3 (IA-terminal) and type 4 (ID-terminal), respectively. The values of D_{max} and p-parameter are indicated. The indices under each value of „p" correspond to the number of a synaptic vesicle recognized as "p-vesicle" in variation succession within individual terminal.

the dLGN more or less evenly. The difference in the distribution of the two above axonal endings is supported by present observations.

The origin of type 3 terminals has not yet been determined precisely, but most likely these endings can be attributed to axons of inhibitory intrageniculate cells which are generally termed I-neurones (GUILLERY, 1969; FAMIGLIETTI and PETERS, 1972; GIOLLI and POPE, 1973; PASIK, PASIK, HAMORI and SZENTÁGOTHAI, 1973).

As it was reported by some investigators the ID-terminals are the derivatives of dendrites of neurones with presynaptic dendrites synapsing on the dendrites of the same or other cells within the dLGN (FAMIGLIETTI and PETERS, 1972; LIEBERMAN and WEBSTER, 1972). Our observations show the postsynaptic position of ID-terminals in relation to the 2a terminals.

The differences among the above three classes of synaptic terminals within the dLGN revealed on the basis of synaptic vesicle size are reliable. These differences correlate with other criteria described previously, such as the size of terminals, the presence or absence of mitochondria, the symmetrical or asymmetrical type of synaptic contact, the localization of the latter on the postsynaptic membrane and the shape of synaptic vesicles. Therefore, our quantitative data support the supposition that the origin of those three classes of presynaptic terminals within the dLGN is different.

The range in the size of 1 and 2b terminals' types overlaps (Brauer and Winkelmann, 1974) and in both cases the distal portions of dendrites serve as a postsynaptic target (Montero and Guillery, 1968; Giolli and Pope, 1973; Brauer et al., 1974), therefore clear differentiation is not possible. The absence of mitochondria cannot be taken as a reliable criterion, because the section place through the terminal is always unknown. Large ranges of both parameters (p and D_{max}) characterizing the vesicle size in terminals of 1 and 2b types, respectively, which is 15—25 per cent for the terminals of the same class and practically overlaps intervals of p and D_{max} distributions for both classes of terminals should be considered. Two interpretations might be offered: 1) non-strict division between the terminals of 1 and 2b types due to mutual mixing of both populations; 2) the existence of some additional classes of synaptic endings which have not yet been discriminated and which are characterized by peculiarities of the ultrastructure similar to those of the terminals of 1 and 2b types (Brückner et al., 1972).

Hence, present data do not allow to come to definite conclusion about true values of parameters p and D_{max} for the terminals of type 1 (cortico-geniculate endings) and 2b (probably retinogeniculate endings). However, both populations of terminals can be differentiated from the other three classes of the dLGN's terminals: the first two have synaptic vesicles of clear round shape of medium size which is between the maximum borders within the dLGN of the albino rat.

Significant difference between the type 2a and 2b terminals, which is based upon the determination of p-parameter and exceeds 8—13 nm or 15—20 per cent, is of special interest for the analysis of neuronal origin of the terminals in the dLGN. This new finding allows to support the idea of heterogeneity of cellular sources of these two types of axon terminals. In view of the hypothesis concerning the retinal origin of these two types of axon endings (Brauer et al., 1974) the occurence of corresponding subpopulations of ganglion cells in the retina may be supposed. This is in agreement with some observations of morphologists and physiologists (Guillery, 1969; Fukuda, 1973; Brauer and Winkelmann, 1977). However, current classifications of the dLGN's terminals might not turn true in details. Some of large terminals (R-boutons) may have non-retinal origin and belong to fibers arriving to the dLGN from other thalamic nuclei or brain stem structures (Brückner et al., 1972). This point will be discussed more extensively in our further paper (in preparation) where the effects of bilateral enucleation on the distribution of various terminal populations in the dLGN of the rat are investigated.

The demonstration of significant differences among at least three classes of synaptic terminals in the dLGN which was based on the statistical analysis of synaptic vesicle size is one of the most important results of the present study. In connection with the hypothesis on close correlation between the synaptic visicle size and molecular nature of the transmitters (Evans, 1966; Lenkov, 1972) this size difference may be interpreted as an indication of different nature of transmitter substances which are released from the synaptic terminals of some classes in the dLGN (cortical, retinal and intrageniculate axon terminals).

Acknowledgements

The authors extend their thanks to Prof. Dietmar Biesold of Paul Flechsig Institute of Brain Research (Leipzig, DDR) for generous encouragement during this investigation and for helpful discussion and comments to the manuscript. We are also grateful to Mrs. Heidrun Kayser, Mrs. Brigitte Matzner and to Mr. Hans-Werner Lungwitz for valuable technical assistance and to Dr. O. P. Uchastkin (Leningrad State University, USSR) for the correction of English translation of the manuscript.

References

Bogolepov, N. N.: Ultrastructure of synapses in the norm and Pathology (1975), Medicine, Moscow (In Russian).

Brauer, K., and E. Winkelmann: Quantitative Charakterisierung der afferenten Axone in der Pars dorsalis des Corpus geniculatum laterale (Cgl d) der Albinoratte an Golgi-Präparaten. Z. mikrosk.-anat. Forsch. **88**, 1110 bis 1124 (1974).

Brauer, K., and E. Winkelmann: Zur Topographie retinogeniculärer Endigungen und Klassifikation von Relais-Zellen im Corpus geniculatum laterale, Pars dorsalis (Cgl d) der Albinoratte in Beziehung zum X- und Y-System. Verh. Anat. Ges. **71**, 973—977 (1977).

Brauer, K., E. Winkelmann, I. Marx and H. David: Licht- und elektronenmikroskopische Untersuchungen an Axonen und Dendriten in der Pars dorsalis des Corpus geniculatum laterale (Cgl d) der Albinoratte. Z. mikrosk.-anat. Forsch. **88**, 596—626 (1974).

BRÜCKNER, G., H. BAER and D. BIESOLD: Zur Ultrastruktur der synaptischen Kontakte im Corpus geniculatum laterale der Ratte. Z. mikrosk.-anat. Forsch. **86**, 513 bis 530 (1972).

CUENOD, M., C. SANDRI and K. AKERT: Enlarged synaptic vesicles as an early sign of secondary degeneration in the optic nerve terminals of the pigeon. Journ. Cell. Sci. **6**, 605–613 (1970).

DEL CERRO, M., and R. S. SNYDER: Studies on the developing cerebellum ultrastructure of the growth cones. Journ. comp. Neurol. **133**, 341–362 (1969).

EVANS, E. M.: On the ultrastructure of the synaptic region of visual receptors in certain vertebrates. Z. Zellforsch. **71**, 499–516 (1966).

FAMIGLIETTI, E. V., and A. PETERS: The synaptic glomeru lus and the intrinsic neuron in the dorsal lateral geniculate nucleus of the cat. Journ. comp. Neurol. **144**, 285 bis 334 (1972).

FROESCH, D.: A simple method to estimate the true diameter of synaptic vesicles. Journ. Microsc. **98**, 85–89 (1973).

FROESCH, D., and R. MARTIN: Heterogeneity of synaptic vesicles in the squid giant fibre system. Brain Res. **43**, 573–579 (1972).

FUKUDA, J.: Differentiation of principal cells of the rat lateral geniculate nucleus into two groups, fast and slow cells. Exp. Brain Res. **17**, 242–260 (1973).

GIOLLI, R. A., and J. E. POPE: The mode of innervation of the dorsal lateral geniculate nucleus and the pulvinar of the rabbit by axons arising from the visual cortex. Journ. comp. Neurol. **147**, 129–144 (1973).

GUILLERY, R. W.: The organization of synaptic interconnections in the laminae of the dorsal lateral geniculate nucleus of the cat. Z. Zellforsch. **96**, 1–38 (1969).

KARNOVSKY, M. J.: A formaldehyde-glutaraldehyde fixative of high osmolarity for use in electron microscopy. Journ. Cell Biol. **27**, 137A–138A (1965).

KAWANA, E., K. AKERT and H. BRUPPACHER: Enlargement of synaptic vesicles as an early sign of terminal degeneration in the rat caudate nucleus. Journ. comp. Neurol. **132**, 297–307 (1971).

KORNELIUSSEN, H.: Vesicular ultrastructure of normal and stimulated motor endplates. Z. Zellforsch. **130**, 28–57 (1972).

LENKOV, D. N.: Vesicular hypothesis of synaptic transmission. In: Rev. Sci. & Techn., Ser. „Human and Animal Physiol.", (Ed. Stepansky, G. A.), **10**, 40–123 (1972) (In Russian).

LENKOV, D. N., T. A. BRAGINA and S. G. MALOSHEVSKY: The synaptic vesicle size in axon terminals of sensory-motor cortex of the cat. Dokl. Akad. Sci. USSR **216**, 427–429 (1974), (In Russian).

LENKOV, D. N., T. A. BRAGINA and S. G. MALOSHEVSKY: Comparison of synaptic vesicle size in axonal profiles synapsing on the neuronal bodies and dendrites in the cat motor cortex. Cytology (Tsitologia) **17**, 1325–1327 (1975) (In Russian).

LIEBERMAN, A. R., and K. E. WEBSTER: Presynaptic dendrites and a distinctive class of synaptic vesicles in the rat dorsal lateral geniculate nucleus. Brain Res. **42**, 196 to 200 (1972).

LIEBERMAN, A. R., and K. E. WEBSTER: Aspects of the synaptic organization of intrinsic neurons in the dorsal lateral geniculate nucleus. An ultrastructural study of the normal and experimentally deafferented nucleus in the rat. Journ. Neurocytol. **3**, 677–710 (1974).

MALOSHEVSKY, S. G., D. N. LENKOV and T. A. BRAGINA: On the changes in synaptic vesicles of axonal terminals of cat stimulated motor cortex. Neurophysiology (Kiev) **7**, 639–646 (1975) (In Russian).

MCKINLAY, R. G., and P. N. R. USHERWOOD: The role of synaptic vesicles in transmission at the insect nerve-muscle junction. Life Sci. **13**, 1051–1056 (1973).

MONTERO, V. M., and R. W. GUILLERY: Degeneration in the dorsal lateral geniculate nucleus of the rat following interruption of the retinal or cortical connections. Journ. comp. Neurol. **134**, 211–242 (1968).

PASIK, P., T. PASIK, J. HÁMORI and J. SZENTÁGOTHAI: Golgi type II interneurons in the neuronal circuit of the monkey lateral geniculate nucleus. Exp. Brain Res. **17**, 18–34 (1973).

VANDER WARDEN, B. L.: The mathematical Statistics. (1960) IL, Moscow, (In Russian translation).

Adresses of authors

Dr. D. N. LENKOV
Physiological Institute
Leningrad University
University Embankment 7/9,
Leningrad, 199164, USSR

Prof. Dr. sc. med. E. WINKELMANN
Dr. rer. nat. K. BRAUER
Dept. of Neuroanatomy
Emilienstr. 14
DDR-701 Leipzig

Dr. rer. nat. G. BRÜCKNER
Dept. of Neurochemistry
Karl-Marx-Städter-Str. 50
Paul-Flechsig-Institute of Brain Research

Buchbesprechung

Architectonics of the Cerebral Cortex

Edit. by M. A. B. BRAZIER and H. PETSCHE.
International Brain Research Organization (IBRO)
Monograph Series. Vol. 3. Raven Press, New York 1978, 486 pp

Der hier vorliegende 3. Band aus der IBRO-Monographie-Serie enthält die Ergebnisse eines Symposiums, das 1976 in Wien anläßlich des 100. Geburtstages Constantin von ECONOMO's stattfand. 29 Beiträge aus den Bereichen der Neuromorphologie, -physiologie und -kybernetik spiegeln den aktuellen Stand dieser Forschungsrichtungen wider.

Große technologische Fortschritte ergaben sich für die Hirnforschung durch die Elektronenmikroskopie, Computeranalysen, Neurochemie, HRP-Markierungen und intrazelluläre Färbetechniken. Die Teilnehmer des Symposiums bemühten sich in den meisten Vorträgen um die Synthese von Struktur und Funktion. Schwierigkeiten der nicht einheitlichen Terminologie ergeben sich sowohl für strukturelle als auch für funktionelle Parameter. Das schmälert jedoch nicht den Wert dieser so sehr notwendigen intensiveren Bemühungen um die gemeinsame Zielstellung. Die Zeiten scheinen vorbei zu sein, in denen der Subspezialist die Frage der Nachbardisziplin mit "that's not my question" abtat!

Von den 29 Beiträgen stammen 13 aus der morphologischen Forschung, 12 aus der Physiologie. 4 Übersichtsreferate geben historische Überblicke mit Summierung des aktuellen Standes zu entsprechenden Fragen. Einleitend würdigt Erna LESKY Constantin von ECONOMO's Leben und Werk. Sein Werk ist **die** Pionierleistung auf dem Gebiet der Cytoarchitektonik, ferner hat er Gesetze der Hirnevolution formuliert und über progressive Cerebralisation geschrieben. Sein Werk war MEYNERT gewidmet, in dessen ehemaligem histologischen Labor von ECONOMO arbeitete.

Mary A. B. BRAZIER referiert über die Architektonik des Cortex cerebri anhand einer historischen Würdigung der wichtigsten Ergebnisse dieser Forschungsrichtung im vorigen Jahrhundert (GERLACH, LEWIS, GOLGI, CAJAL, WEIGERT, NISSL, APÁTHY, LISTER, HODGKIN, SCHWANN, PURKYNĚ, KÖLLIKER, OBERSTEINER, REMAK, DEITERS, HIS, FOREL, HELMHOLTZ). Das 3. morphologische Übersichtsreferat von FLEISCHHAUER gibt einen geschichtlichen Rückblick auf die letzten 50 Jahre und führt zur Vorstellung heutiger Probleme. Die wichtigsten Meilensteine in der Erforschung der corticalen Architektonik aus jüngster Vergangenheit werden in klarer, logischer Wertung hintereinander gereiht: Gründe für sachliche Diskrepanzen zwischen ECONOMO und den VOGTS, spätere Forderungen nach klarer Definition cytoarchitektonischer Kriterien und nach Beachtung individueller Variabilität, die Atlanten von Primatengehirnen (BAILEY, BONIN), LORENTE de No's GOLGI-Studien, die Problematik der Verbindungen zum und vom Thalamus für die architektonische Abgrenzung corticaler Bezirke (ROSE, WOOLSEY), moderne autoradiographische Nachweisverfahren für cerebrale Verbindungen, elektrophysiologische Einzelableitungen, die Spezifität der Neurone für besondere Stimuli, die zur Schlußfolgerung einer mosaikähnlichen Repräsentation in Form vertikaler Zylinder führt. Das Interesse an der horizontalen Cortexschichtung wird heutzutage erweitert durch verstärkte Zuwendung zur Problematik der vertikalen Organisation. Hierzu werden bisherige Ergebnisse und offene Fragen formuliert. Der Beitrag von DETZER bringt Ergebnisse semiautomatischer Erfassung der vertikalen Dendritenbündelung.

Aus der Reihe der morphologischen Beiträge erscheinen einige besonders grundsätzliche und bemerkenswert. PALAY beweist anhand der Meynert-Zelle, einer Sonderform corticaler L V-Pyramiden, die laminäre Zuordnung des Spine-Bestandes an den apikalen Hauptdendriten. Die SCHEIBELS analysieren qualitativ und quantitativ die BETZsche Pyramidenzelle des Menschen, ihre Beteiligung an der dendritischen Bündelung sowie — in Korrelation mit elektrophysiologischen Ergebnissen — ihre funktionellen Möglichkeiten. Anhand sehr schöner GOLGI-Bilder gibt Terez TÖMBÖL eine Klassifikation der Interneurone bei Katze und Kaninchen in unterschiedlichen Cortexregionen. Die Notwendigkeit von Neurontypisierungen besteht zweifellos, gleichzeitig wird vor Überschätzung als alleinige Arbeitsmethode gewarnt. SZENTÁGOTHAI referiert über Systeme corticaler Verbindungen, ihre Spezifität, Selektivität und geometrische Ausdehnung. Almut SCHÜZ berichtet über die funktionelle Wertung dendritischer Spines im Lernprozeß. Der Beitrag von BRAAK bringt sehr schöne neue Ergebnisse der Pigmentarchitektonik für die Erkennung, laminäre Zuordnung von Stern- und Pyramidenzellen sowie für die Bedeutung dieser Verteilungsmuster. WOLFF untersuchte bei der Ratte die Ontogenese corticaler Lamination. FARKAS-BARGETON und DIEBLER haben die neocorticale Enzymreifung (5 Enzyme) an menschlichen Gehirnen prä- und postnatal in verschiedenen Areae verfolgt. Die laminäre Reihenfolge dieser Reifung ist ähnlich wie bei der cytologischen Entwicklung und Migration. Die lange postnatale Reifungsphase deutet auf eine ebenfalls lange Periode der Plastizität. RELAY-Zellen im visuellen Cortex der Katze untersucht D. SANIDES mit Hilfe von HRP-Markierungen.

Ebenfalls am visuellen System der Katze wurden experimentelle Ergebnisse von GAREY vorgestellt: es existieren 2 Zelltypen in Retina und Cgl mit wahrscheinlich unterschiedlicher funktioneller Nuancierung bei Projektion von Erregungen zur Area 17 und 18.

In einem Übersichtsreferat legt CREUTZFELDT grundsätzliche Gedanken dar zur Verschaltung im Neocortex unter dem Aspekt der Korrelation von Struktur und Funktion. Hier wird ein eindrucksvoller Versuch unternommen, bisherige morphologische und physiologische Forschungsergebnisse in grundsätzlichen Bau- und Funktionsprinzipien zu summieren und die Definition der Aufgaben des Neocortex im Sinne eines Filters für Umweltstimuli innerhalb des ZNS zu begründen. Im Beitrag von ADRIANOV findet man ähnliche Gedanken zur corticalen Integration: adaptive Mechanismen werden als wesentliche Vorgänge gesehen, die auch durch die Übereinstimmung von Struktur und Funktion gewährleistet werden.

Die Vielzahl physiologischer Beiträge bringt hauptsächlich elektro-physiologische experimentelle Ergebnisse, wobei meistens versucht wird, die Frage nach der Übereinstimmung von histologischer Struktur und Widerspiegelung bioelektrischer Phänomene so exakt wie z. Zt. möglich zu beantworten. Wie schwierig und technisch begrenzt diese Möglichkeiten noch sind, bringt der Vortrag von PETSCHE am besten zum Ausdruck.

Am Schluß des Bandes gibt der Beitrag von BRAITENBERG eine Übersicht über architektonische Bauprinzipien mit instruktiven quantitativen Daten aus eigenen Untersuchungen und Vergleichen mit der Literatur. Der Autor gelangt zu Modellvorstellungen des Cortex, weist aber berechtigterweise darauf hin, wie wenig bisher die Variabilität corticaler Strukturen quantitativ erfaßt und in diese Gedanken einbezogen werden konnte.

Die Ausstattung des Buches ist hervorragend. Die Kürze der Zeitspanne zwischen Symposium und Erscheinen des vorliegenden Bandes ist wohltuend und dem Anliegen aktueller Information angemessen!

E. SCHULZ, Berlin

J. Hirnforsch. 19 (1978) 423–432

Instituto Cajal, C.S.I.C. Velázquez, 144 - Madrid-6 Spain

Evidences for the existence of glycoproteins and mucopolysaccharides metabolic cycles in the cerebellum of birds and mammals

By

Luis M. GARCIA-SEGURA, R. MARTINEZ-RODRIGUEZ, E. BOGONEZ, R. MARTINEZ-MURILLO and A. TOLEDANO

With 11 figures and 2 tables

(Received 2nd August 1977)

Summary: Cerebellum glycoproteins of rats and ducks are studied using sugar-specific agglutinins. It is used the Con A-Peroxidase-DAB technique described by BERNHARD and AVRAMEAS (1971). Other agglutinins (Tetragonolobus Purpureus Lectin, Wheat Germ Agglutinin and Soybean Agglutinin) fixed to Peroxidase according to the technique of GONATAS and AVRAMEAS (1973) are also used. The techniques for mucopolysaccharides as described MARTÍNEZ-RODRIGUEZ et al., (1976) are carried out as well. By means of all these techniques, results that suggest the existence of metabolic cycles in Purkinje cells are obtained.

The existence of several staining states among neurons of the same kind has been observed since last century and the classic distinction between chromophilic neurons (dark neurons) and chromophobic neurons (light neurons) has also been made (CAJAL, 1909). Morphological variations, generally related to variations of the neuronal size, were described together with differences in the grade of staining. Since then, it was stated to discover whether those differences of staining corresponded to something real or to a fixation artefact. The observation that those differences tock place in neurons which were located closed in the same histological section, lead some authors such as CAJAL (1909) to consider that the effects of fixation might not be the origin of these differences since it was supposed that the fixative might have acted on the two neurons in the same way, being these neurons situated at the same level and therefore the fixative penetration was the same for both. Recently it has been demonstrated that the existence of neurons with different coloration characteristics can be revealed by specific histochemical methods. Thus, several staining states have been distinguished by using techniques for proteins (DIXON, 1953, 1954, 1958), lipids (KROON, 1963; KROON and GOOSENS, 1967; TZARNOUCHANOV, 1967; ROZEMOND, 1970; GARCÍA-SEGURA et al., 1974, 1976a) and mucopolysaccharides (DIXON and HERBERTSON, 1950; RODRÍGUEZ-PÉREZ, 1958; MARTÍNEZ-RODRÍGUEZ et al., 1975, 1976a, b; GARCÍA-SEGURA et al., 1976b, 1977a.) These results suggest the existence of alternate functional states. While some authors continue ascribing to the neuronal death due to fixation defects the staining differences among neurons, others have obtained results that support the idea that these differences are the expression of metabolic cycles. Thus, in dorsal root ganglia, alternate states of condensation and pulverization of Nissl bodies, leaving to different metachromatic states, have been distinguished (GARCÍA-SEGURA et al., 1977). Similar results have been obtained in the sympathetic ganglia (GARCÍA-SEGURA et al., 1976b) and these results coincide with the presence of neurons with different electronic density in these ganglia (HOSSACK and WYBURN, 1954; HESS, 1955; DAWSON et al., 1955; ANDRES, 1961; TENNYSON, 1965; BUNGE et al., 1967). Therefore, there are electron microscopy and histochemical evidences (TEWARI and BOURNE, 1962) for the existence of metabolic cycles in the neurons of dorsal root and sympathetic ganglia. The presence of electron dense neurons in the Nervous System of mammals (see for instance ROGER et al., 1972) transfer the old light microscopy plannings to the electron microscopy level.

In the present work the glycoproteins of cerebellum are studied by using specific plant agglutinings and the cerebellar mucopolysaccharides by means of adecuate histochemical techniques. Attention is payed to Purkinje cells, in which results are observed that suggest the existence of production and consumption cycles of proteins and mucopolysaccharides.

Material and Methods

A) Experimentation animals:

Ducks (Anas platyrynchus, L) a week old and male rats (Wistar strain) a month old were used. The animals were perfused with 10% formaldehyde in PBS and the cerebellums were post-fixed in the same fixative for 24 hours at 4 °C. Then the cerebellum was dehydrated and embedded in paraffin.

B) Techniques for proteins

The agglutinins used were the following: Concanavalin A (Con A), Tetragonolobus purpureus Lectin (TPA), Wheat Germ Agglutinin (WGA) and Soybean Agglutinin (SBA) Type VI. All these agglutinins were purchased from Sigma. Their purity was tested by electrophoresis and their activity by agglutination test. It was used the technique of BERNHARD and AVRAMEAS (1971) in order to detect the binding sites of Concanavalin A to the tissue. Binding of Concanavalin A to peroxidase (Horseradish Peroxidase, Sigma, Type II) was studied by agar immunodifussion tests and quantitative precipitation in liquid (GARCÍA-SEGURA et al., 1977b). Rest of agglutinins were fixed to peroxidase by the technique of GONATAS and AVRAMEAS (1973). For this purpose, a peroxidase which had active aldehyde groups, was prepared by treating this enzime with glutaraldehyde for 24 hours at 22 °C. Glutaraldehyde excess was removed by using a Sephadex G-25 column (AVREMEAS and TERNYNCK, 1971). The agglutinins were made react with active peroxidase by methods described by GONATAS and AVRAMEAS (1973).

Preparations that were not treated with agglutinin, treated only with peroxidiase (normal and active), treated only with the DAB solution and treated only with the agglutinin and DAB without inserting peroxidase, were used as control preparations. Controls preparations, in which endogenous peroxidases were inhibited by the technique of STRAUSS (1971), were also used.

C) Techniques for mucopolysaccharides

The techniques and controls carried out, were described previously (MARTÍNEZ-RODRÍGUEZ et al., 1976), (PAS, Alcian Blue, Colloidal Iron).

Results

Positivity was observed in the cytoplasm of Purkinje cells in the duck cerebellar sections treated with Con A. This positivity was more intense in some neurons than in others, in such a way that some neurons appeared stained with rather intensity while in others the staining was very slight and in others absolutely negative (Fig. 7). Positivity was observed in the basket-like networks. In a lot of slightly positive neurons the reaction appeared only located either in the cell membrane area or in the perinuclear area (Fig. 7). Similar results were seen in the rat cerebellum. In the cerebellum sections of duck treated with WGA fixed to peroxidase, positivity was observed in the cytoplasm of some Purkinje cells. The granule cells presented a very tempered positivity (See table II). In some Purkinje

Fig. 1, 2 and 3. Binding of Wheat Germ Agglutinin to Purkinje neurons of duck cerebellum. In Fig. 1 and 3 various coloration states are observed in Purkinje cells: in some of them positivity appeared all over the cytoplasm, in others only in the perinuclear area. In addition cytoplasm positively is variable.

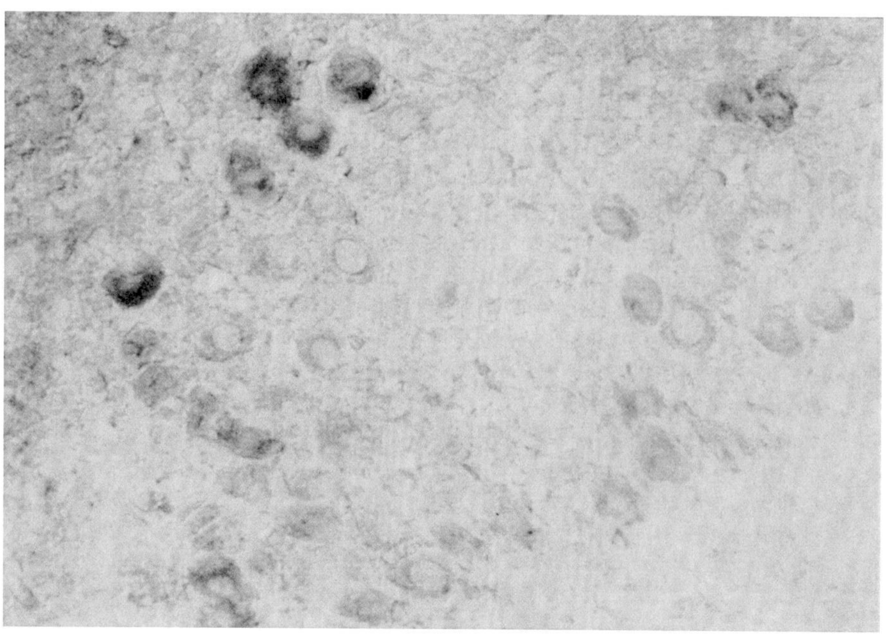

Fig. 1

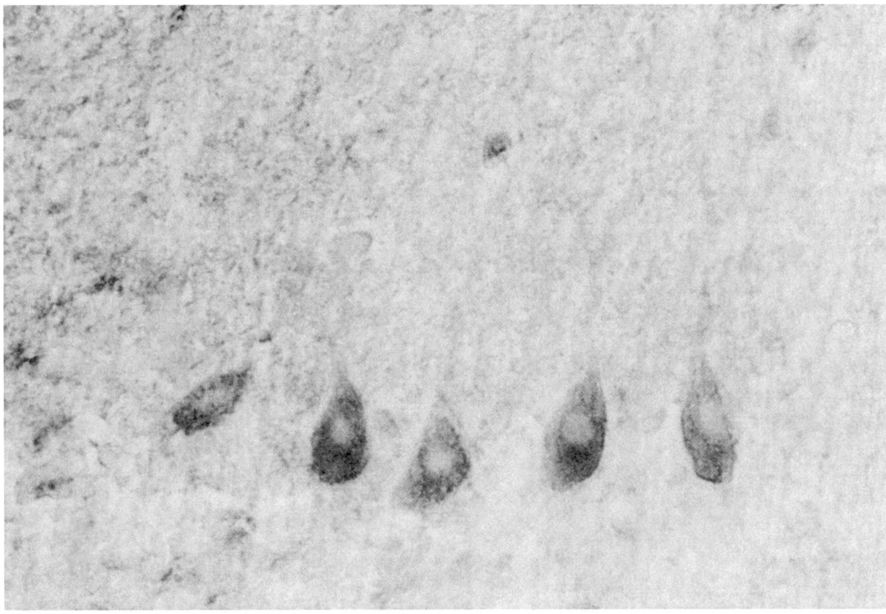

Fig. 2

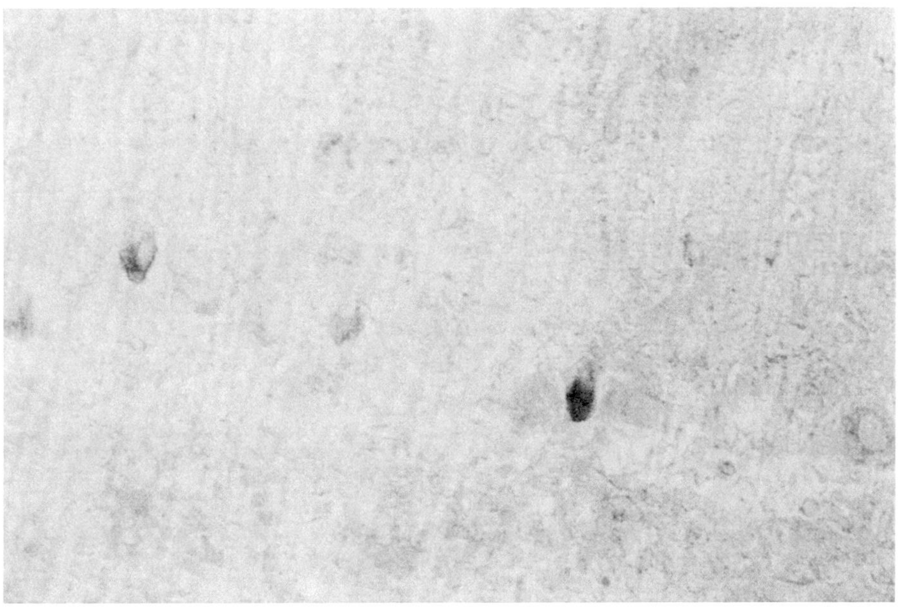

Fig. 3

Table 1. Staining of Purkinje Cells after the treatment with PAS, Alcian Blue and Colloidal Iron techniques. Assayed agglutinins binding to Purkinje cells (duck and rat). +++: highly positive, +: positive, −: negative.

	ConA	WGA	TPA	SBA	PAS	Collodal Iron	Alcian Blue
Nucleolus	−	−	−	−	−	+	+
Nucleus	−	−	−	−	−	−	−
Perinuclear area	+	+	+	+	+	+	+
Cytoplasm	+	+	+	+	+	+	+
Plasmal membrane	+	+	+	+	+	+++	+++

Table II. Binding to the Purkinje cells layer and granular layer of sugar-specific agglutinins.
+: strong binding, +/−: weak binding.

	Duck	RAT
Purkinje cells	+	+
Granular cells	+/−	+

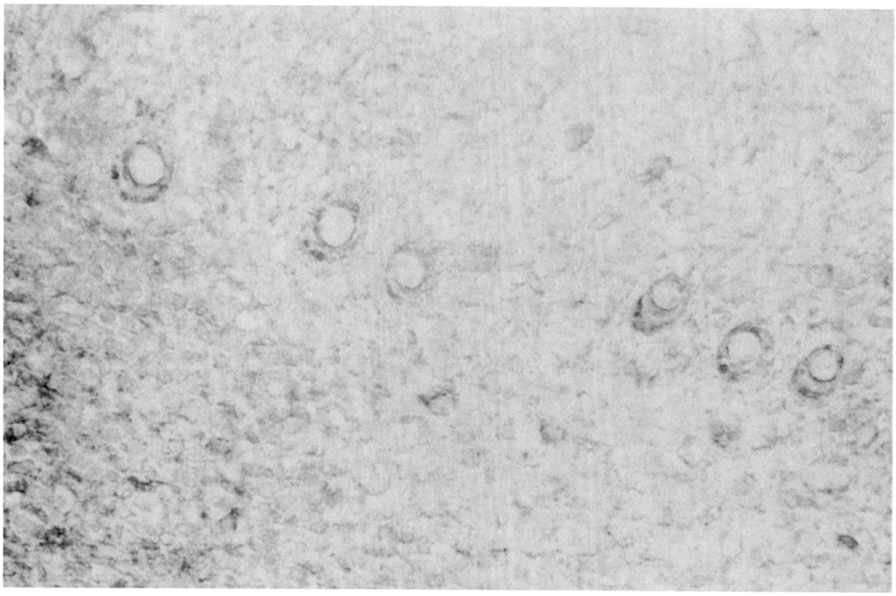

Fig. 4

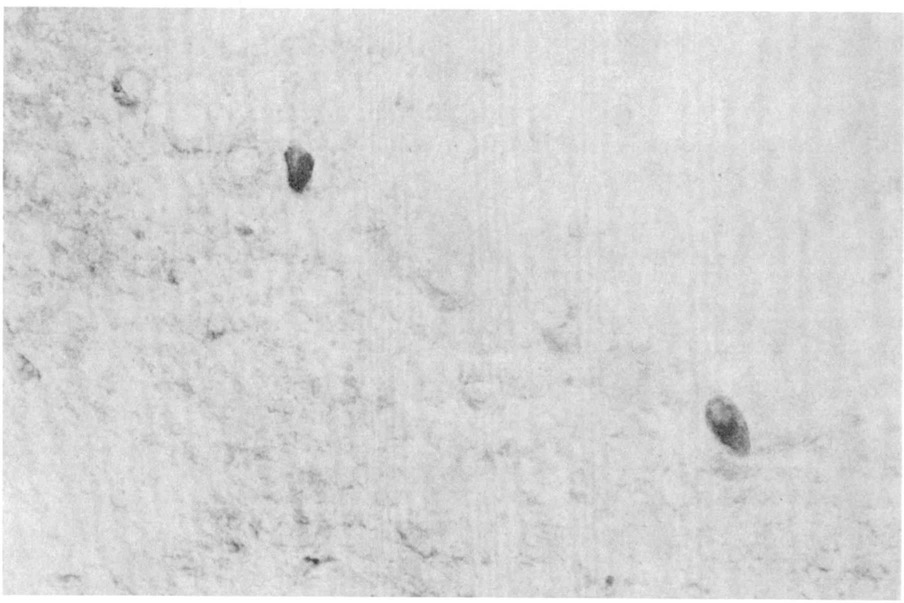

Fig. 5

Fig. 4 to 6. Binding of different agglutinins to Purkinje cells of duck. 4: TPA. 5 and 6: SBA. Observe the existence of neurons that have captured a great amount of agglutinin (dark neurons) while others, located in the vecinity, presents a weak positivity or are negative. In Fig. 4 positivity is preferentelly located in the perinuclear area and in the cytoplasm periphery.

cells the only area that presented positivity was the perinuclear area. In the rat cerebellum treated with WGA fixed to peroxidase, positivity was observed in the cytoplasm, perinuclear area and cell membrane of Purkinje cells. Most of Purkinje cells presented positivity and only some of them were negative. Granular layer presented a lesser positivity than Purkinje cells but this positivity was more intense than that of the duck granular layer (See table II).

Cerebellar sections treated with SBA and TPA fixed to peroxidase, showed very similar results to that observed with WGA (Fig. 1—6). Results are summarized in table I.

Cerebellar sections of duck stained with the PAS method showed some Purkinje cells with a very positive cytoplasm while in others the cytoplasm was weakly stained or was negative. In some Purkinje cells, positivity was only observed in the nucleolus, cell membrane and perinuclear area, however the cytoplasm was negative or presented some positive granulations. In the sections treated with the Colloidal Iron or with Alcian Blue methods, Purkinje

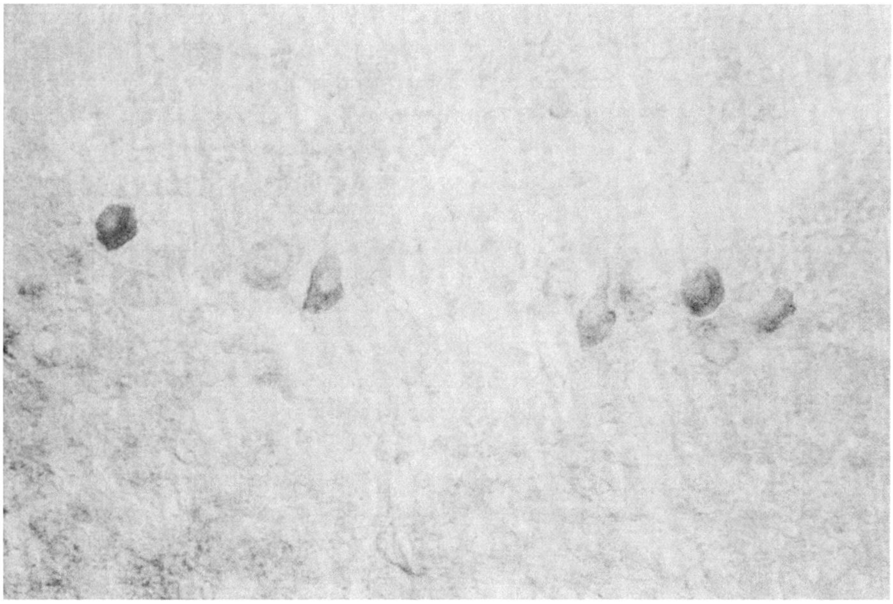

Fig. 6

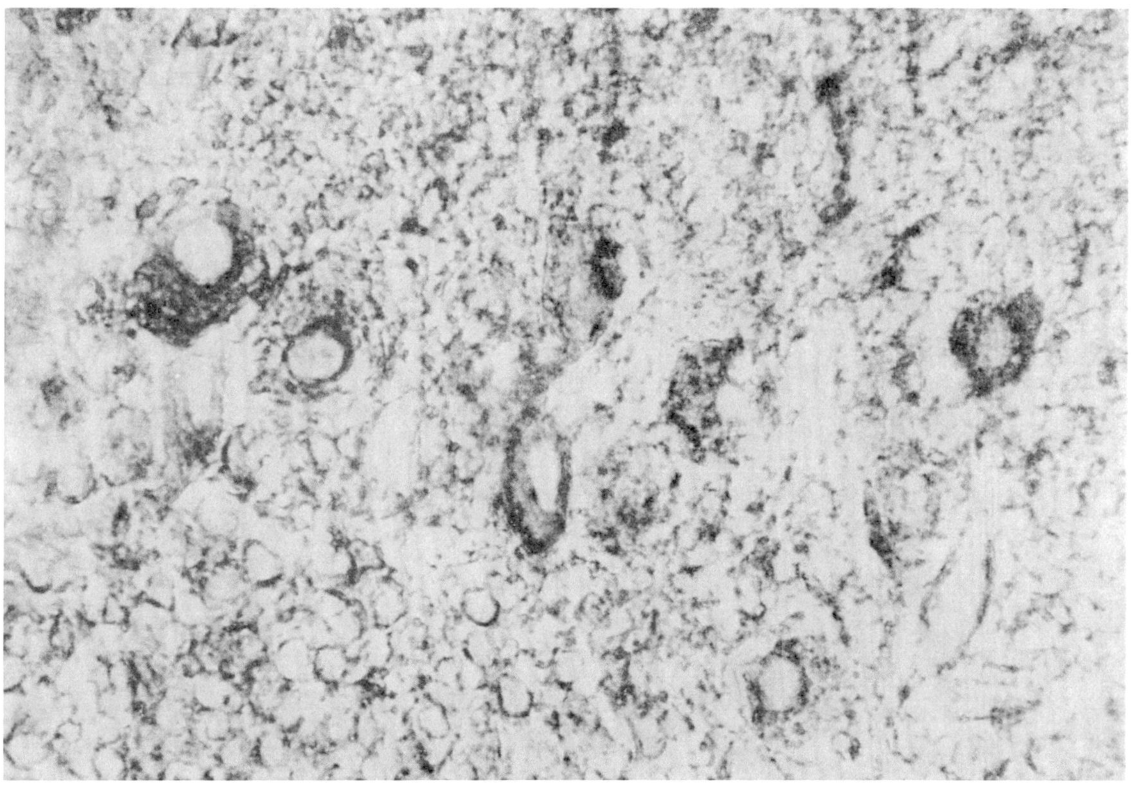

Fig. 7. Binding of Concanavalin A to Purkinje cells of duck cerebellum. In some neurons positivity is located on plasma membrane whereas in others it is presented in the perinuclear area or all over the cytoplasma.

cells showing different staining states, were observed. Some of them presented the tinction in the cytoplasm specially in the perinuclear area and cell membrane. This cytoplasm positivity was of different intensity from some neurons to others. In some neurons the cytoplasm appeared negative but the cell membrane appeared highly positive (Fig. 10 an 11). Results obtained with Con A, WGA, SBA, and TPA were not modified by the endogenous peroxidases inhibition. Controls were negative except when the treatment with agglutinin was omitted and the section was treated with peroxidase and then with the DAB

Fig. 8

solution. In this cases, nonespecific uptake of peroxidases by the nucleolus was observed. The nucleoli of Purkinje cells, granular cells as well as of Golgi cells and another neurons (central cerebellar nuclei neurons) and glia cells appeared positive. This positivity resists prolonged wahsing with PBS and phosphate buffer at pH 6.2 (Palladini and Solheid, 1976) (Fig. 8 and 9). The surprising fact of these results is that when the sections were previously treated with any of the agglutinins at concentrations above 0.5 mg/ml, the nucleoli did not appear positive. On the other hand the nonespecific uptake of peroxidase was limited to the nucleolus (sometimes to the whole nucleus) while the cytoplasm appeared negative.

The staining of the plasma membrane with Alcian Blue and Colloidal Iron decreases after prior metilation but resists the hyaluronidase, beta-glucuronidase, sulphatase and ribonuclease actions. On the contrary, the staining of the cytoplasm is weaker to the enzimatic extraction.

Discussion

In the Purkinje cells of rat and duck cerebellum, staining differences have been found by means of techniques for lipids (García-Segura et al., 1974; Meitner, 1977) and a techniqne of very complex specificity: the tannin-iron method (García-Segura,

Fig. 8, 9. Non-specific binding of paroxidase to nucleolus (See text).

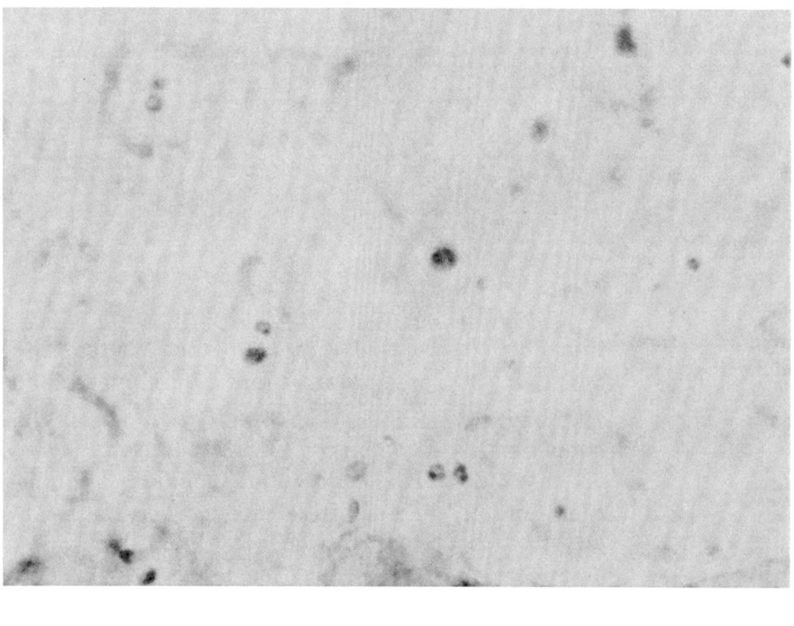

Fig. 9

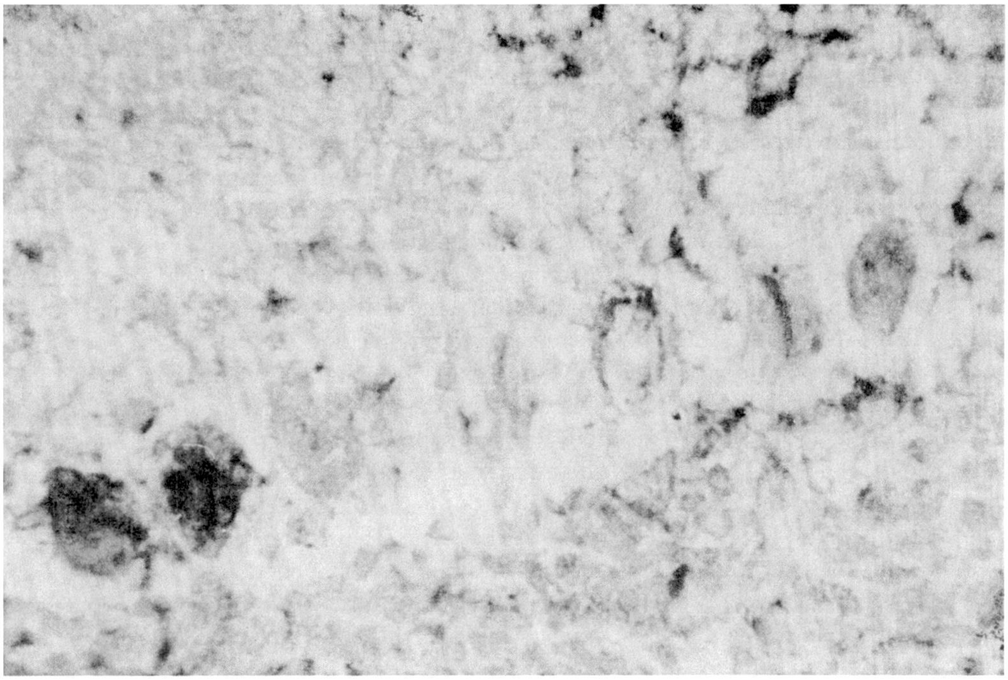

Fig. 10

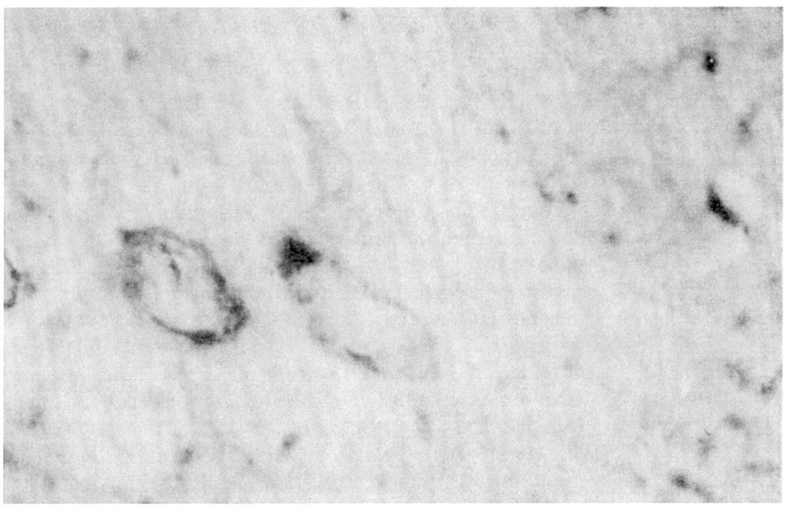

Fig. 11

Fig. 10 and 11. Colloidal Iron reaction. Negative neurons are observed. In others the cytoplasm presents an important positivity. Notice that some neurons (in Fig. 11 the two neurons on the left) present a positivity being located exclusively on the plasma membrane (Purkinje cells of duck cerebellum).

1974a, b; García-Segura and Díaz-González, 1975). Results obtained using last method suggest the existence of a functional cycle similar to that evidenced in the dorsal root and sympathetic ganglia by techniques for mucosubstances (García-Segura et al., 1976, 1977a). Evidences for the existence of metabolic cycles in Purkinje cells of mammals have also been found by using the distribution of several enzimatic activities (Tewari and Bourne, 1962b, c, d, 1963a, b, c; Iijima and Nakajima 1964; Shanta et al., 1967; Shantaveerappa and Bourne, 1965; Iijima, 1977). Golgi apparatus of Purkinje cells shows different fragmentation states in rat (Iijima, 1977) as well as in many other types of neurons. This cyclic activity of Golgi apparatus may be related to the different staining states observed in the present work when plant agglutinins were used. In a study under electron microscope Wood et al. (1974) found that Convanavalin A bound to Golgi apparatus of Purkinje cells and as well the endoplasmic reticulum membranes showed Con A binding sites (Wood et al., 1974; Wood and McLaughlin, 1976). Thus, the different grades of staining found by us unde light microscope may be expression of different amounts of endoplasmic reticulum and Golgi apparatus, that is different biosynthetic activities.

Concanavalin A reacts specifically with glycoproteins which contain branched terminal nonreducing alpha-D-glycopyranosyl, alpha-D-mannopyranosyl or beta-D-fructofuranosyl residues (Goldstein et al., 1965, 1969; Agrawal and Goldstein, 1972). Wheat Germ Agglutinin presents specificity to N-actyl-glucosamine residues (Sharon and Lis, 1972; Lis and Sharon, 1973; Marchesi, 1972). Tetragonolobus purpureus Lectin presents affinity to alpha-L-fucose residues (Sharon and Lis, 1972; Lis and Sharon, 1973; Yariv et al., 1972). Soybean Agglutinin presents affinity to N-acetyl-D-galactosamine and D-galatose residues (Sharon and Lis, 1972; Lis and Sharon, 1972, 1973). Our results demonstrate therefore that in the Purkinje cells of rat and duck cerebellum exist glycoproteins containing branched terminal alpha-L-fucose, N-acetyl-D-galactosamine, N-acetyl-glucosamine and alpha-D-glucopyranosyl and/or alpha-D-mannopyranosyl and/or beta-D-fructofuranosyl residues. Moreover, these residues present different distribution from some neurons to others and their presence and concentration also vary. These facts indicate the existence of a cyclic activity in Purkinje cells which is showed by means of a cycle of glycoproteins production and consumption (or transport out of the perikarya).

In the central cerebellar nuclei of the rat, two types of staining location have been found by means of the Alcian Blue and Colloidal Iron Methods (Martínez-Rodríguez et al., 1975). Some neurons present positivity only in the edge of cytoplasm whereas others showed positivity all over the cytoplasm. In the present work we have observed these two staining types in Purkinje cells of the rat and duck. Results obtained with these techniques and with the PAS method, in the cytoplasm of Purkinje cells, suggest the existence of a cytoplasmic mucopolysaccharides cycle that perhaps is also related to the cycle of the Golgi apparatus. On the other hand, the cellular edge staining is also more intense in some neurons than in others. This fact suggests a consumption and replacement of the mucosubstances located in that area.

Results obtained in the present work suggest the existence of metabolic cycles of Purkinje cells glycoproteins and mucopolysaccharides. This fact indicates that it is possible to differenciate neurons with various activity states by means of histochemical methods. Recently a technique allowing to visualize the glucuse consumption in sections of nervous system has been developped (Sokoloff et al., 1974; Sokoloff, 1975; Kennedy et al., 1975a; Sokoloff et al., 1977). This technique that is founded in the accumulation of 2-deoxy-D(^{14}C)glucose-6-phosphate in the tissues after the injection of 2-deoxy-D-(^{14}C)-glucose, allows to observe regions where a functional activity exists as a result of stimulation (Kennedy et al., 1975a, b, 1976). The results obtained with this technique indicate that it is possible to visualize modifications of the neuronal acitivity. On the other hand, other studies have demonstrated that histochemical alterations in the enzimatic activities (ATPase, acid phosphatase, G-6PDH, MDH) are evidenced in the motor spinal neurons of animals forced to run and then sacrificed (Yellin, 1967; Grechman et al., 1975). All these results offer new possibilities to histochemical correlations of the neuronal activity. The existence of metabolic cycles for glycoproteins and mucopolysaccharides in Purkinje cells of the birds and mammals cerebellum demonstrated in this work, is an additional evidence for the different neuronal states of activity that can be observed by histochemical techniques.

References

Agrawal, B. B. L., and I. J. Goldstein: Concanavalin A, the Jack Bean (Canavalia ensiformis) Phytohemagglutinin. Methods in Enzymol., **28**, 313–318 (1972).

Andres, K. H.: Untersuchungen über den Feinbau von Spinalganglien. Z. Zellforsch., **55**, 1–49 (1961).

Bernhard, W., and S. Avrameas: Ultrastructural visualization of cellular carbohydrate components by means of concavalin A. Exp. Cell. Res., **64**, 232–236 (1971).

Bunge, M. B., R. P. Bunge, E. R. Peterson and M. R. Murray: A light and electron microscope study of long term organized cultures of rat dorsal root ganglia. J. Cell. Biol., **32**, 439–466 (1967).

Cajal, S. R.: Histologie du système nerveux de l'homme et des vertébrés. Vol. I, pp. 209–214, Paris Maloine, 1909 (Edition 1972: Instituto Cajal, Madrid).

Gammermeyer, J.: Histochemical phospholipid reaction in ischemic neurons as an indication of exposure to postmortem trauma. Exp. Neurol., **49**, 252–271 (1975a).

Cammermeyer, J.: The effect of postmortem trauma on neuronal cell types stained histochemically for phospholipids. Exp. Neurol., **46**, 616–633 (1975b).

Dawson, I. M., J. Hossack and G. M. Wyburn: Observations on the Nissl's substance, cytoplasmic filaments and the nuclear membrane of spinal ganglion cells. Proc. Roy. Soc. (London) Ser. B, **144**, 132 (1955).

Dixon, K. C.: Glycolysis and cytochemistry of cerebral cortex. J. Physiol., Lond., **120**, 267–277 (1953).

Dixon, K. C.: Cytochemistry of cerebral gray matter. Q. Jl. exp. Physiol., **120**, 267–277 (1954).

Dixon, K. C.: Differentiation of chromophilic and chromophobic neurons. J. Anat., **92**, 425–432 (1958).

Dixon, K. C., and B. M. Herbertson: A cytoplasmic constituent of brain. J. Physiol. (Lond.), **111**, 244–247 (1950).

García-Segura, L. M.: Utilización del método tano-férrico

para el estudio del Sistema Nervioso Central al microscopio óptico. Aplicación en el estudio de los núcleos cocleares y vestibulares y sus conexiones en el pollo. Memoria de Licenciatura. Universidad Complutense de Madrid, Facultad de Ciencias. Madrid, 1974a.

García-Segura, L. M.: The use of tannin-iron method as a nervous system general stain technique. Trab. Inst. Cajal Inv. Biol., **66**, 253—256 (1974b).

García-Segura, L. M., and P. Díaz-Gonzalez: Etude sur la spécificité histochemique de la méthode tanno-ferrique. Ann. Histochim., **20**, 153—158 (1975).

García-Segura, L. M., R. Martinez-Rodriguez, A. Toledano, M. de Agustín and C. Rodriguez-Gonzalez: Investigación histoquímica de fosfolípidos y de ésteres triglicéridos simples en cerebelo de rata. Trab. Inst. Cajal inv. biol., **66**, 235—245 (1974).

García-Segura, L. N., R. Martinez-Rodriguez, A. Toledano and M. Beltran: Lipids distribution in dog spinal cord: a histochemical study. An. Anat., **25**, 75—84 (1976a).

García-Ssegura, L. M., R. Martinez-Rodriguez and A. Toledano: A histochemical investigation on mucopolysaccharides in the dog sympathetic ganglia. Acta Histochem., **56**, 66—72 (1976b).

García-Segura, L. M., R. Martinez-Rodriguez and A. Toledano: Mucopolysaccharides in the dog spinal cord and dorsal root ganglia. Acta Anat., **98**, 91—96 (1977a).

García-Segura, L. M., R. Martinez-Murillo, E. Bogonez, A. Toledano, C. Rodriguez-Gonzalez and R. Martinez-Rodriguez: Glicoproteinas, mucopolisacaridos y polianiones en las neuronas de Purkinje del cerebelo del pato. Trab. Inst. Cajal inv. Biol., in press (1977b).

Gerchman, LeRoy, B., V. R. Edgerton and R. E. Carrow: Effects of physical training on the histochemistry and morphology of ventral motor neurons. Exp. Neurol., **49**, 790—801 (1975).

Goldstein, I. J., C. E. Hollerman and E. E. Smith: Protein-carbohydrate interaction. II. Inhibition studies in the interaction on Concanavalin A with polysaccharides. Biochemistry, **4**, 876—883 (1965).

Goldstein, I. J., L. L. So, Y. Yang and Q. C. Callies: Proteincarbohydrate interaction XIX. The interaction of Concanavalin A with IgM and the Glycoprotein Phytohemagglutinins of the Waxbean and the Soybean. J. Immunol., **103**, 695—698 (1969).

Gonatas, N. K., and S. Avrameas: Detection of plasma membrane carbohydrates with lectin peroxidase conjugates. J. Cell. Biol., **59**, 436—443 (1973).

Hess, A.: The fine structure of young and old spinal ganglia. Anat. Rec., **123**, 399—424 (1955).

Hossack, J., and G. M. Wyburn: Electron microscopic studies of spinal ganglion cells. Proc. Roy. Soc. Edinburgh B. **65**, 239 (1954).

Iijima, K.: Histochemical studies on the morphology of the Golgi apparatus and on the distribution of some enzymes concerned with carbohydrate metabolism in the rat cerebellum. Acta. Histochem., **58**, 163—186 (1977).

Iijima, K., and Y. Nakajima: Histochemical studies on the oxidative enzymes of the human cerebellum Bull. Tokyo Med. Dent. Univ., **11**, 103—135 (1964).

Kennedy, C., M. H. des Rodiers, J. W. Jehle, M. Reivich, F. Sharpe and L. Sokoloff: Mapping of functional neural pathways by autoradiographic survey of local metabolic rate with (^{14}C) deoxyglucose. Science, **187**, 850—853 (1975a).

Kennedy, C., M. des Rosiers and L. Sokoloff: The ocular dominance columns of the striate cortex as studies by the deoxyglucose method for measurement of local corebral glucose utilization. Trans. Am. Neurol. Ass., **100**, 74—77 (1975b).

Kennedy, C., M. H. des Rosiers, O. Sakurada, M. Shinohara, N. Reivich, J. V. Jehle and L. Sokoloff: Metabolic mapping of the primary visual system of the monkey by means of the autoradiographic (^{14}C) deoxyglucose technique. Proc. Nat. Acad. Sci., U.S.A., **73**, 4230—4234 (1976).

Kroon, D. B.: Certain cells in the hypothalamic neurosecretory nuclei which are stainable by the acid-haemateintest for phospholipids according to Baker. Z. Zellforsch., **61**, 317—337 (1963).

Kroon, D. B., and E. M. L. D. Gossens: The demonstration of phospholipids in certain cells in neurosecretory nuclei in the rat with Baker's acid haematein after fixation in glutaraldehydeformol-calcium. Z. Zellforsch., **83**, 527—582 (1967).

Lis, H., and N. Sharon: Soy Bean (Glycine max) Agglutinin. In: Methods in Enzymol., **28**, 360—365 (1972).

Lis, H., and N. Sharon: The biochemistry of plant lectins (phytohemagglutinins). Ann. Rev. Biochem., **42**, 541—574 (1973).

Marchesi, V. T.: Wheat germ (Triticum vulgaris) agglutinin. In: Methods in Enzymol., **28**, 354—356 (1972).

Martinez-Rodriguez, R., L. M. García-Segura, A. Toledano, M. Gonzalez, A. Gamonal, M. de Agustín, P. Díaz-Gonzalez and C. Rodriguez-Gonzalez: Investigation of acid mucosubstances in rat's cerebellum. Ann. Histochim., **20**, 205—215 (1975).

Martinez-Rodriguez, R., A. Toledano, L. M. García-Segura and C. Rodriguez-Gonzalez: Histochemical study of mucopolysaccharides in the subthalamic region of rats. Acta Histochem., **56**, 200—210 (1976a).

Martinez-Rodriguez, R., Al Toledano, L. M. García-Segura, M. Gonzalez, A. Gamonal, P. Díaz-Gonzalez, M. de Agustín and C. Rodriguez-Gonzalez: Mucopolysaccharides in hypothalamic neurons of the rat. J. Anat., **121**, 231—239 (1976b).

Meitner, E. R.: Unterschiede in der Färbbarkeit der Purkinje-Zellen Acta Anat., **97**, 191—199 (1977).

Palladini, G., and Solheid, C.: Absorption of free and antibodyconjugated horseradish peroxidase to fixed glial cells in tissue culture, a possible source of error in immunochemistry. Acta Histochem., **57**, 87—92 (1976).

Pineda, A., D. S. Maxwell and L. Kruger: The fine structure of neurons and satellite cells in the trigeminal ganglion of cat and monkey. Am. J. Anat., **121**, 461—488 (1967).

Rodriguez-Perez, A. P.: Estudios histoquimicos en el sistema nervioso periférico. II. Trab. Inst. Cajal inv. Biol., **50**, 23—31 (1958).

Roger, J. F., J. R. Holbrook and R. S. Pozos: Electron dense cells in the nervous system of representative mammals. J. Submicrosc. Cytol., **4**, 135—159 (1972).

Roozemond, R. C.: Phospholipid composition of some parts of rat hypothalamus and its relation with some histochemical observations. J. Neurochem., **17**, 179—188 (1970).

Shantha, T. R., K. Iijima and G. Bourne: Histochemical studies on the cerebellum of squirrel monkey (Saimiri sciuretus), Acta Histochem., **27**, 129—162 (1967).

Shanthaveerappa, T. R., and G. H. Bourne: The thiamine pyrophosphatase technique as an indicator of the morphology of the Golgi apparatus in the neurons IV.

Studies on the cerebellum of the rat and squirrel monkey. Z. Zellforsch., **68**, 699—719 (1965).

SHARON, N., and H. LIS: Lectins: Cell-agglutinating and sugar-specific proteins. Science, **177**, 949—959 (1972).

SOKOLOFF, L.: The (^{14}C) deoxyglucose method for the quantitative determination of local cerebral glucose utilization. In: Blood Flow and Metabolism in the Brain, A. M. Harper, W. B. Jennett, J. D. Miller and J. O. Rowan (Eds.), Churchill Livingstone, Edinburgh, pp. 15. 1.—15. 8. 1975.

SOKOLOFF, L., M. REIVICH, C. S. PATLAK, K. D. PETTIGREW, M. DES ROSIERS and C. KENNEDY: The (^{14}C) deoxyglucose method for the quantitative determination of local cerebral glucose consumption. Trans. Am. Soc. Neurochem., **5**, 85 (1974).

SOKOLOFF, L., M. REIVICH, C. KENNEDY, M. H. DES ROSIERS, C. S. PATLAK, K. D. PETTIGREW, O. SAKURADA and M. SHINOHARA: The (^{14}C) deoxyglucose method for the measurement of local cerebral glucose utilization: theory, procedure, and normal values in the conscious and anesthetized albino rat. J. Neurochem., **28**, 897—916 (1977).

STRAUS, W.: Inhibition of peroxidase by methanol and by methanol-nitroferricyanide for use in immunoperoxidase procedures. J. Histochem. Cytochem., **19**, 682—688 (1971).

TENNYSON, V. M.: Electron microscope study of the developing neuroblast of the dorsal root ganglion of the rabbit embryo. J. Comp. Neurol., **124**, 267—318 (1965).

TEWARI, H. B., and G. H. BOURNE: Histochemical evidence of metabolic cycles in spinal ganglion cells of the rat. J. Histochem. Cytochem., **10**, 42—64 (1962a).

TEWARI, H. B., and G. H. BOURNE: Histochemical distribution of enzymes associated with the metabolism of carbohydrates in the various elements of the cerebellum. Exp. Cell. Res., **28**, 444—449 (1962b).

TEWARI, H. B., and G. H. BOURNE: Histochemical studies on the distribution of oxidative enzymes in the cerebellum of the rat. J. Histochem. Cytochem., **10**, 619—627 (1962c).

TEWARI, H. B., and G. H. BOURNE: Histochemical studies on the distribution of specific and non-specific cholinesterases and simple esterase in the cerebellum of the rat. Acta anat., **51**, 349—368 (1962d).

TEWARI, H. B., and G. H. BOURNE: Histochemical studies on the localization of adenosine triphosphate in the cerebellum of the rat. J. Histochem. Cytochem., **11**, 246—257 (1963a).

TEWARI, H. B., and G. H. BOURNE: Histochemical studies on the distribution of alkaline and acid phosphatases and 5'-nucleotidase in the cerebellum of the rat. J. Anat., **97**, 65—72 (1963b).

TEWARI, H. B., and G. H. BOURNE: Histochemical studies on the distribution of monoamine oxidase in the cerebellum of the rat. Acta anat., **52**, 334—340 (1963c).

TZARNOUCHANOV, P. K.: Sur l'histochimie des lipides du corps du neurone. Recherches sur les neurones de la moelle épinière du rat blanc. Ann. Histochim., **12**, 83—90 (1967).

WOOD, J. G., and B. J. MCLAUGHLIN: Cytochemical studies of lectin binding sites in smooth membrane cisternae of rat brain. Brain Res. **118**, 15—26 (1976).

WOOD, J. G., B. J. MCLAUGHLIN and R. P. BARBER: The visualization of Concanavalin A binding sites in Purkinje cell somata and dendrites of cat cerebellum. J. Cell. Biol., **63**, 541—549 (1974).

YARIV, J., A. J. KALB and S. BLUMBERG: Lotus tetragonolobus L-Fucose-binding proteins. Methods in Enzymol., **28**, 356—360 (1972).

YELLIN, H.: Neural regulation of enzymes in muscle fibers of red and white muscle. Exp. Neurol., **19**, 92—103 (1967).

The authors adress:

Dr. Luis M. GARCIA-SEGURA
Instituto Cajal, C.S.I.C.
Velázquez, 144 - Madrid-6
Spain

J. Hirnforsch. 19 (1978) 433–437

From the Department of Normal Anatomy, Institute of Biostructure, Academy of Medicine in Poznań
(Direktor: Prof. dr. hab. med. W. Woźniak)

Dentritic range of the neurons of the intermediate gray at the levels of the first and second lumbar segments of the spinal cord in the cat.
3. The range of dendrites of the lateral region neurons

By

Kazimierz Grottel and Teresa Hofman

With 3 figures

(Received September 2, 1977)

Summary: The first and second lumbar segments of the spinal cord in kittens and mature cats were examined with Golgi method.
It was found that the dendrites of large and medium neurons situated in the lateral region of the intermediate gray penetrated lamina V dorsally, and lamina VIII and motor nuclei ventrally. Laterally these dendrites reached the lateral horns and even the lateral funicles. Paracentrally they ran along the whole intermediate gray of the cord, entered the thoracic nucleus, and even the white commissure.

Zusammenfassung: Mit Hilfe der Golgi-Methode wurden das 1. und 2. Lumbalsegment des Rückenmarkes junger und erwachsener Katzen untersucht. Es wurde gefunden, daß die Dendriten der großen und mittleren Neurone, die in der lateralen Region der Pars intermedialis (grisea) gelegen sind, dorsal in die Lamina V und ventral in die Lamina VIII und die motorischen Kerne eindringen. Lateral erreichen diese Dendriten die Seitenhörner und sogar die Seitenstränge. Parazentral verlaufen sie durch die ganze Pras intermedia (grisea) des Rückenmarkes, treten in den Nucleus thoracicus und sogar in die Commissura alba. Die Ergebnisse werden mit Angaben des Schrifttums verglichen, die sich auf morphologische und physiologische Untersuchungen stützen.

In the previous investigations (Grottel and Hofman; Grottel and Hofman, in print) we found that the neurons of the central and paramedial regions of the intermediate gray at the level of the first and second lumbar segments differed not only in size, as had been suggested in earlier reports based on Nissl's stained sections (Rexed, 1952, 1954), but also in the range of their dendrites. This refers not only to the neurons of neuclei described earleir (intermediomedial nucleus, thoracic nucleus or nuclei of the paracommissural cells pointed out by us) but also to individual cells like the cells of the central region of the intermediate gray. The size of neurons and various range of dendrites of particular groups of neurons suggest the different activity of the neurons in these regions.

The present studies are based on the observation of the dendritic range of neurons situated in the lateral region of the intermediate substance at the first and second lumbar segments of the spinal cord. The division of the intermediate substance was shown in the diagrams of previous papers (Grottel and Hofman; Grottel and Hofman,in print), as well as in figures 1, 2, and 3 of the paper.

Material and methods

The first and second lumbar segments of 8 mature cats of both sexes, and 4 four days old, 4 seven days old and 3 twelve weeks old kittens were examined.

One segment from each group kittens was stained according to Nissl's method, the others were impregnated according to rapid Golgi and Bubenaite methods (Matsushita, 1969). Both segments of two mature cats, impregnated according to Golgi method, were cut in sagittal plane, and of two other cats in horizontal plane. In specimens stained with toluidine blue the sections were of 10–20 μm thick, and in those impregnated according to Golgi method the sections were of 100–250 μm thick. The Nissl's method was used mainly to determine the localization of individual cellular groups and to compare our results with those of other authors. The examinations of the range of dendrites were based on Golgi method and its modifications. The graphic reconstructions were the same as described previously (Grottel and Hofman, in print; Grottel 1976, 1977).

Results

Among the neurons of the lateral region of the intermediate gray at the investigated levels the medium size cells (13–15 × 15–20 μm) prevail. These are

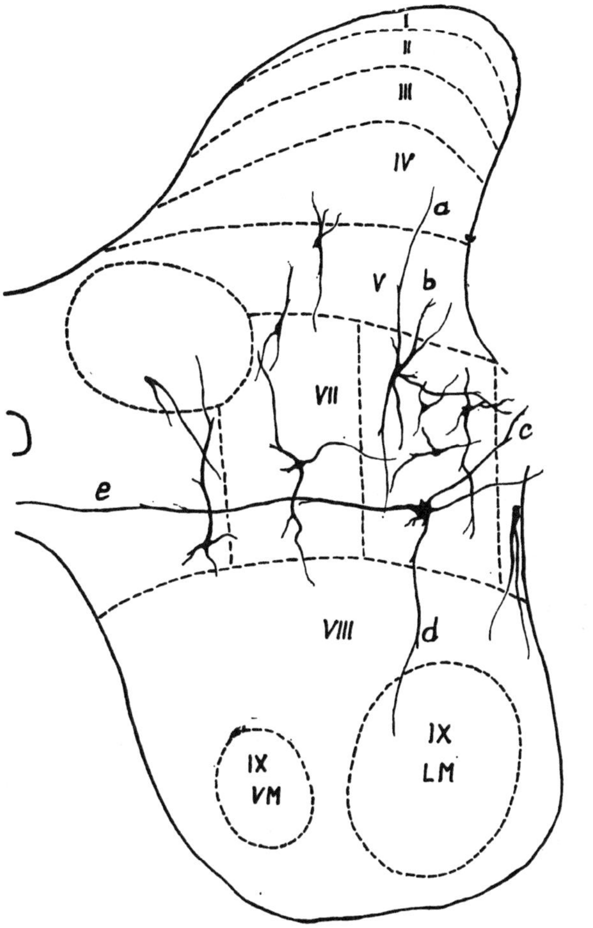

Fig. 1. Semischematic figure showing transverse section of spinal cord on left side at L_1 level. Division into laminae is based on Rexed's description (1952, 1954). Figure was drawn from one section of 200 μm thick taken from seven day old kitten. Magn. ×80.

characteristic for Rexed's lamina VII. The large neurons found in the region are of greater diameters than described by Rexed (40 × 50 μm). Dendrites of small cells (8—10 × 10 μm) of the lateral region of the spinal intermediate gray branch off in all directions for relatively short distances. In some instances, however, the course of dendrites is limited mainly to one plane, e.g. to the transverse plane of the cord.

The arrangement of dendrites of medium and large neurons lying in this region is mainly dorsoventral in transverse sections. They ramify mainly within the lateral region of the intermediate gray. The cells which are oriented dorsally penetrate most frequently the adjoining part of lamina V (Figs. 1a, 2c, 3c), and these oriented verntally send most of their dendrites to the lamina VIII (Fig. 2h, j, l). However, some dorsally located cells of the region give off long dendrites to the lamina VIII even (Fig. 3e) whereas a number of ventrally lying neurons give off dendrites to the Lamina V (Fig. 2d). It was also found that certain medium and large neurons of the lateral region lying medially, laterally, dorsally as well as ventrally send their dendrites both into the central region (Figs. 2g, 3d) of the intermediate gray and lateral horn (Figs. 1c, 2a, 3a) as well.

The longest dendrites of the medium and large neurons of the spinal intermediate gray reach the lamina IX ventrally (Figs. 1d, 2i, k, 3f).

In a few cases the large neurons of the lateral region send their longest dendrites to the white commissure. They pass medially and ventrally in the transverse section of the spinal cord. Such dendrites running medially and dorsally penetrate the thoracic nucleus (Fig. 2e). Some of dendrites passing ventrally to the thoracic nucleus traverse the central and paramedial regions of the intermediate gray (Fig. 2f) and they are seen in the posterior commissure.

Laterally, medium and large neurons give off dendrites to lateral horn. These frequently enter the lateral funiculus (Figs. 2b, m, 3b). Generally their dorsal range is not wide for they do not pass beyond the lamina V. In some cases, however, medium and

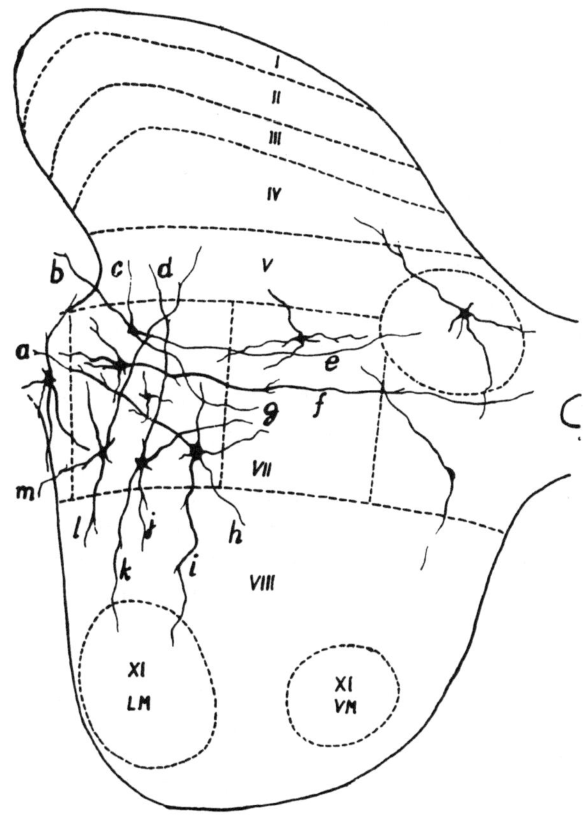

Fig. 2. Semischematic figure showing transverse section of spinal cord on right side at L_2 level. Figure was drawn from one section of 250 μm thick taken from nature cat. Magn. ×80

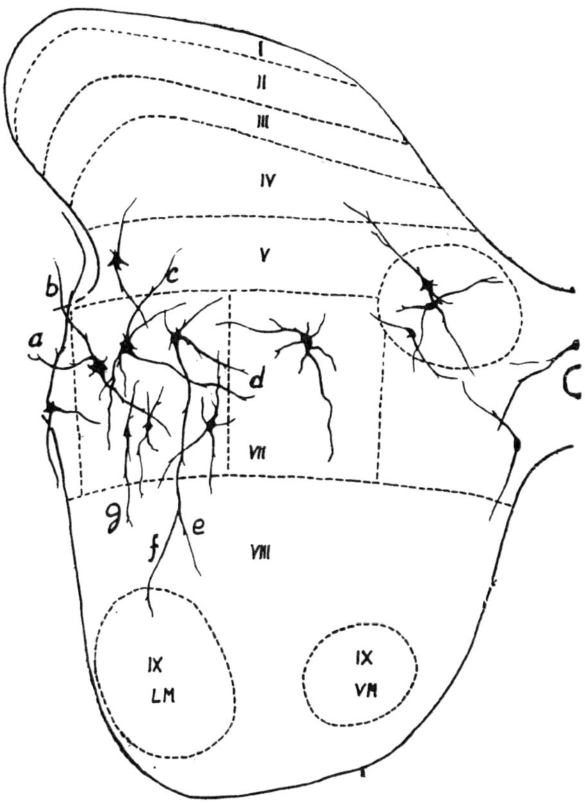

Fig. 3. Semischematic figure showing transverse section of spinal cord on right side at L_2 level. Figure was drawn from one section 200 µm thick taken from mature cat. Magn. × 80.

large neurons send their dendrites to the lamina IV (Fig. 1a).

The largest neurons of the lateral region of the intermediate gray have the longest dendrites. This can be seen best in the horizontal and sagittal sections of the spinal cord where the differences in the lengths of dendrites are very clear as the longest dendrites of all neurons (especially of large and medium ones) usually run along the axis of the cord.

Discussion

The description of dorsal and ventral ranges of dendrites has been mainly based on the study of transverse sections of the cord for we have followed the laminar pattern of REXED (1952, 1954). Although in sagittal sections the course and division of dendrites directed into both horns is clearly seen, it is difficult to discern the boundaries of the individual laminae of the cord. Besides, the position of some nuclei may be doubtful as with the Golgi method some of the neurons are unstained (SCHEIBEL and SCHEIBEL, 1971). The horizontal and sagittal sections, however, illustrate the division of dendrites in these planes and show well the range of the longest dendrites of most neurons situated in axis of the cord.

The basic difference between the cells of the lateral and paramedial regions of the spinal intermediate gray is in the range of the dendrites especially of the medium and large cells. The occurrence of the intermediomedial nucleus and the paracommissural cells (GROTTEL and HOFMAN, in print) within the paramedial region makes also the difference. In the central region the dendrites of large neurons frequently reach the lamina III. Not only the cells situated dorsally in this region, but also many of these situated more ventrally (GROTTEL and HOFMAN, in print) have the same orientation of dendrites. The dendrites of neurons of the lateral region very rarely ramify dorsally beyond the lamina V.

According to NYBERG-HANSEN (1966) the corticospinal fibers and rubrospinal tracts terminate mainly in the lateral parts of laminae V and VII. Lamina VIII and ventral parts of lamina VII are terminal points for the fibers of the lateral vestibulospinal tract of the same side of the cord, for medial vestibulospinal, interstitiospinal and for some reticulospinal tracts of both sides. A similar location (lateral parts of laminae VI and VII) has been found by KOSTYUK and SKIBO (1975) for some rubrospinal fibers. KUYPERS and BRINKMAN (1970) confirmed the termination fo the corticospinal tracts suggested by NYBERG-HANSEN.

BALDISSERA and BRUGGENCATE (1976), using physiologic methods, have found that the contralateral rubrospinal tract terminate on some VSCT neurons. Monosynaptic terminations of the tract evoke the EPSPs whereas bisynaptic endings produce IPSPs. The remaining VSCT cells of this region are stimulated or inhibited from the rubrospinal tract by the way of interneurons. The rubrospinal tract facilitates bisynaptic IPSPs in VSCT interneurons in the pathway of Ia and Ib inhibitory interneurons, both from flexors and extensors. Transmission to Ia interneurons is a bisynaptic and to Ib interneurons is a monosynaptic. Since, as BALDISSERA and BRUGGENCATE (1976) found, bisynaptic EPSPs and IPSPs were transmitted by the rubrospinal tract both in contra- and ipsilateral cutaneous and high treshold muscular afferents, we presume that the ipsilateral transmission could have been twofold. Firstly it reached the ipsilateral interneurons which sent their dendrites to the other side of the cord or secondly to the contralateral interneurons via the white commissure. The contralateral interneurons affected the dendrites of ipsilateral VSCT cells which sent their dendrites to the other side of the cord. We noticed such neurons in the lateral region of the intermediate gray, though they seemed more scarce there than in the central or paracentral regions of the cord. This hypothesis seems to be even more justified since Kostyuk and Skibo (1975) found that the rubrospinal tract terminated mainly on peripheral parts of dendrites of neurons, and on their branches. It must be remembered that the rubrospinal tract runs in the spinal cord contralaterally to the position of the red nucleus. The cells mediating afferentats transmitted to the other side of the cord, according to Oscarsson (1964), lie in the dorsal horn so their participation would not have been taken into account in the examinations of Baldisserra and Bruggencate. Even though the endings of axons of the corticospinal, rubrospinal and other supraspinal tracts mentioned above, at least on some neurons of the investigated region exist, we assume that the neurons may receive still other afferents which have not been examined. The cells of the region in question, sending their dendrites into the paramedial part of the intermediate gray of the cord into the ventral horns, to the lamina V and even, via the white commissure, to the other side of the cord, may receive, by the way of the dendrites, afferents from the intraspinal tracts as well as dorsal root afferents. Therefore, it is possible that they are connected with, for instance, reticulospinal medullary fibers, which according to Nyberg-Hansen (1966) terminate in laminae VIII and IX.

Similarly, the results obtained by Sterling and Kuypers (1967) proving that the fibers of dorsal roots do not reach lateral parts of laminae V and VI, do not deny that at least some cells of the investigated region, sending their dendrites to the thoracic nucleus or to ventral horns, receive afferents from the dorsal roots. The latest reports of Hancock, Rigamonti and Bryan (1973) and Hancock, Foreman and Willis (1975) prove that the cells of the region examined by us receive both cutaneous and visceral afferents. It is possible for the spatial arrangement of these neurons, according to the above authors, is very wide.

As we know, Sprague (1953), Oscarsson (1964), Gustafsson and Lindström (1970), Burke, Lundberg and Weight (1971), localize the cell bodies of the ventral spinocerebellar tract also in the lateral part of lamina VII. According to Jankowska and Lindström (1972) inhibitory interneurons in Ia pathway are scattered among these cells.

Our results presenting the range of dendrites of neurons of this region, considering the suggestions of the physiological studies, may provide the morphological explanation of the neuronal connections. These connections have been also found in the examinations of Lindström (1973), Hultborn, Illert and Santini (1976 a, c, b) and other authors who applied physiological methods.

References

Baldissera, F. and G. Bruggencate: Rubrospinal effects on ventral spinocerebellar tract neurons. Acta physiol. scand. **96**, 233–249 (1976).

Burke, R., A. Lundberg and F. Wieght: Spinal border cell origin of the ventral spinocerebellar tract. Exp. Brain Res. **12**, 283–294 (1971).

Grottel, K.: Dendrite bundles of the ventral horn cells of the thoracic portion of the spinal cord in cats. Folia Morphol. (Warsz.), **35**, 393–403 (1976).

Grottel, K.: Distribution of dendrites in the ventral horn of the thoracic part of the spinal cord in the sagittal plane in cats. Folia Morphol. (Warsz.), **36**, 29–36 (1977).

Grottel, K. and T. Hofman: Dendritic range of the neurons of the intermediate gray at the levels of the first and second lumber segment of the spinal cord in the cat. I. The range of dendrites of the central region neurons (in print).

Grottel, K. and T. Hofman: Dendritic range of the neurons of the intermediate gray at the levels of the first and second lumber segment of the spinal cord in the cat. II. The range of dendrites of the paramedial region neurons (in print).

Gustafsson, B. and S. Lindström: Depression of Ia IPSP in spinal border cells by impulses in recurrent motor axon collaterals. Acta physiol. scand., **80**, 4, 13A (1970).

Hancock, M. B., D. D. Rigamoti and R. N. Bryan: Convergence in the lumbar spinal cord of pathways activated by splanchnic nerve hind limb cutaneous nerve stimulation. Exp. Neurol. **38**, 337–348 (1973).

Hancock, M. B., R. D. Foreman and W. D. Willis: Convergence of visceral and cutaneous input onto spinothalamic tract cells in the thoracic spinal cord of the cat. Exp. Neurol., **47**, 240–248 (1975).

Hultborn, H., M. Illert and M. Santini: Convergence on interneurones mediating the reciprocal Ia inhibition of motoneurones. I. Disynaptic Ia inhibition of Ia inhibitory interneurones. Acta physiol. scand., **96**, 193–201 (1976).

Hultborn, H., M. Illert and M. Santini: Convergence on interneurones mediating the reciprocal Ia inhibition of motoneurones. II. Effect from segmental flexor reflex pathways. Acta physiol. scand., **96**, 351–367 (1976).

Hultborn, H., M. Illert and M. Santini: Convergence on interneurones mediating the reciprocal Ia inhibition of motoneurones. III. Effect from supraspinal pathways. Acta physiol. scand., **96**, 368–391 (1976).

Jankowska, E. and S. Lindström: Morphology of interneurones mediating Ia reciprocal inhibition of motoneurones on the spinal cord of the cat. J. Physiol. **226**, 805–823 (1972).

Kostyuk, P. G. and G. G. Skibo: Synaptic connections of the rubrospinal tract fibers in the cat spinal cord. Neurol. and Neusurg., **39**, 5, 309 (1975).

Lindström, S.: Recurrent control from motor axon collaterals of Ia inhibitory pathways in the spinal cord of the cat. Acta physiol. scand., suppl. 392 (1973).

Matsushita, M.: Some ascpects of the interneuronal connections in cats spinal gray matter. Comp. Neurol., **136**, 57–80 (1969).

NYBERG-HANSEN, R.: Functional organization of descending supraspinal fibre systems to the spinal cord. Anatomical observations and physiological correlations. Reviews of Anatomy, Embryology and Cell Biology, Springer-Verlag, Berlin, Heidelberg, New York (1966).

OSCARSSON, O.: Differential course and organization of uncrossed and crossed long ascending spinal tracts. Progress in Brain Research, Editors: J. C. Eccles and J. P. Schadè, Physiology of spinal neurons, **12**, 164—178 (1964).

REXED, B.: The cytoarchitectonic organization of the spinal cord in the cat. J. Comp. Neurol., **96**, 415—466 (1952).

REXED, B.: A cytoarchitectonic atlas of the spinal cord the cat. J. Comp. Neurol., **100**, 297—380 (1954).

SCHEIBEL, M. E. and A. B. SCHEIBEL: Inhibition and the Renshaw cell. A structural critique. Brain Behav. Evol. **4**, 53—93 (1971).

SPRAGUE, J. M.: Sinal ,,border cells'' and their role in postural mechanism (Schiff-Sherrington phenomenon). J. Neurophysiol., **16**, 464 (1953).

STERLING, P. and H. G. J. M. KUYPERS: Anatomical organization of the brachial spinal cord of the cat. I. The distribution of dorsal root fibers. Brain Res. **4**, 1, 1—15 (1967).

The authors Adress:

Dept. of Normal Anatomy Institute of Biostructure,
Academy of Medicine in Poznań,
Av. Swięcickiego 6, Poland

Buchbesprechung

Afferent and Intrinsic Organization of Laminated Structures in the Brain

(7th International Neurobiology Meeting)
Editor: CREUTZFELD, O., Göttigen, Germany
Experimental Brain Research/Supplement 1
127 figs. XXIII, 579 pages. 1976. Soft cover DM 58,–; US $ 25.60
Reduced price for subscribers to the journal "Experimental Brain Research"
Soft cover DM 40,–; US $ 17.60
Berlin–Heidelberg–New York: Springer Verlag
ISBN 3-540-07923-8

Der vorliegende Band enthält die Vorträge des „7th International Neurobiology Meeting" vom September 1975 in Göttingen und stellt eine repräsentative Dokumentation des gegenwärtigen Standes wissenschaftlicher Auffassungen über die funktionelle Organisation verschiedener corticaler Strukturen des Gehirns dar. Die spezielle Zielstellung dieses Kongresses bestand in der Analyse unseres gegenwärtigen Wissens über die morpho-funktionellen Grundlagen laminärer Strukturen des Cerebellums, von Allo- und Neocortex sowie des Mittelhirns. Ausgehend von klassischen und modernen neuroanatomischen und physiologischen Arbeiten, schließen die Beiträge Aspekte der Entwicklungs-Neurobiologie, Neurochemie und theoretischen Neurobiologie ein. Unterschiedliche Cortices haben als Basisprinzip ihrer inneren Organisation laminäre Strukturen, dem die Neurone mit ihren Dendritenfeldern, Axonen und Afferenzen folgen, wobei als Basisprinzip der afferenten Organisation eine zumeist strenge somatotopische Organisation der spezifischen Inputs gilt.

Gerade das Kapitel „Cerebellum" vermittelt deutlich, welche Fortschritte auf dem Gebiet der Hirnforschung durch Anwendung komplexer Untersuchungsmethoden erreicht werden. In den Beiträgen von ECCLES, MUGNAINI und CHAN-PALAY werden die morphologischen und physiologischen Grundlagen von der neuronalen Struktur über die inneren Verbindungen des neuronalen Netzwerkes bis zu durch Transmitternachweis charakterisierte Bahnen gelegt. Experimentelle Modelle (ALTMAN, HAMORI, SOTELO, WOODWARD) – wie z. B. Röntgenbestrahlung in bestimmten Entwicklungsstadien oder Untersuchungen mit Mutanten – erscheinen besonders geeignet, Probleme der Differenzierungs- und Reifungsmechanismen, Induktionswirkungen einzelner Neuronpopulationen oder afferenter Systeme und der synaptischen Reorganisation zu klären. Wichtige Hinweise zur Funktion des Cerebellums geben auch die Untersuchungen zur Aufklärung der afferenten und efferenten Kleinhirnverbindungen mit elektrophysiologischen, autoradiographischen, Tracer- und Degenerationsmethoden.

Die Beiträge im Kapitel „Allocotex" stellen eine wesentliche, vorwiegend funktionell orientierte Ergänzung zu dem kürzlich von STEPHAN veröffentlichten Handbuchband „Der Allocortex" dar. So beinhalten die Vorträge von BLACKSTAD, HJORT-SIMONSON, HAUG, SHIPLEY, ZIMMER und LYNCH neben morphologischen Grundlagen der laminären Struktur der Hippocampusformation neue Ergebnisse über die inneren Verbindungen, über Entwicklungsprozesse, plastische Veränderungen nach Deafferenzierung und zu putativen Transmittern (STORM-MATHISEN). Die klar gegliederte laminäre Struktur des Hippocampus mit der „bandförmig" angeordneten synaptischen Organisation bietet nicht nur für Physiologen ein ideales In vivo-Untersuchungsobjekt (vgl. ANDERSEN), sondern dient als In vitro-Modell („tissue slice technique", Explantat-Kultur) für komplexe Untersuchungen unter definierten Umweltbedingungen – unabhängig von Veränderungen der respiratorischen und zirkulatorischen Pulsation, des Blutdruckes und anderer exogener bzw. endogener Umweltfaktoren. Neue morphologische, biochemische und elektrophysiologische Ergebnisse belegen das eindrucksvoll.

In einer größeren Anzahl von Vorträgen über den Neocortex werden neuere Ergebnisse und Hypothesen über Entwicklung, corticale Kreise und die funktionelle Organisation der Hirnrinde mitgeteilt (vgl. RICHMAN, RAKIC, SEIL, WOLFF, CAVINESS). Studien zur Gliadifferenzierung und Neuronentwicklung zeigen die Fortschritte in der Aufklärung von Mechanismen und Faktoren, der plastischen Eigenschaften des ZNS und die Notwendigkeit kombinierter quantitativ-morphologischer und neurophysiologischer Untersuchungsmethoden.

Ausgehend vom Konzept des „basic circuitry" des Neocortex (SZENTÁGOTHAI), der laminären und „columnären" Organisationim visuellen System, stehen Fragen der Korrelation von morphologischen Veränderungen und funktioneller Beeinflussung, von funktionellen Eigenschaften der rezeptiven Felder und der synaptischen Konnektivität bis zu den Grundlagen der Gedächtnisprozesse im Vordergrund (HUBEL, HESS, ROVAMO).

Der letzte Teil des Bandes beinhaltet die Vorträge über subcorticale, laminäre Strukturen des Mittelhirns, vorwiegend von Anteilen des visuellen Systems sowie des Tectum opticum bei verschiedenen Tierspezies.

Abschließend bleibt festzustellen, daß der vorliegende Band in eindrucksvoller Weise nicht nur den Stand, sondern auch Fortschritte und Trends in den Neurowissenschaften widerspiegelt.

J. WENZEL, Berlin

J. Hirnforsch. 19 (1978) 439–443

From the Department of Normal Anatomy, Institute of Biostructure, Academy of Medicine in Poznan
(Director: Prof. dr. hab. med. W. Woźniak)

Dentritic range of the neurons of the intermediate gray at the levels of the first and second lumbar segments of the spinal cord in the cat. 4. The range of dendrites of the lateral horn neurons

By

Kazimierz Grottel and Teresa Hofman

With 2 figures

(Received September 2, 1977)

Summary: The topography of dendrites of the lateral horn neurons at the level of L_1-L_2 have been investigated in young and mature cats using Golgi impregnation method.
It has been stated that the dendrites of investigated neurons reach the lamina III (dorsally), the lateral funiculi (laterally), and the lateral and central parts of the intermediate gray (medially). They penetrate the thoracic nucleus as well. In their ventral course they were found within the laminae VIII and IX. Afferent connections of the neurons are also discussed.

Zusammenfassung: Bei juvenilen und adulten Katzen wurde die Topographie der Dendriten von Neuronen des Cornu lateralis des Rückenmarkes in Höhe von L 1 und L 2 nach Anwendung der Golgi-Methode untersucht. Es wurde festgestellt, daß die Dendriten der untersuchten Neurone dorsal die Lamina III, lateral die Seitenstränge und medial die lateralen und zentralen Anteile der Pars intermedia (grisea) erreichen. Gleichfalls dringen sie in den Nucleus thoracicus ein. Ventral verlaufen sie bis in die Laminae VIII und IX. Die afferenten Verbindungen dieser Neurone werden diskutiert.

Our previous papers dealt with the direction of dendritic course of the neurons within the central (Grottel and Hofman, in print), paramedial (Grottel and Hofman, in print) and lateral regions (Grottel and Hofman, in print) of the intermediate gray. It was pointed out that the connections of the neurons could be established far away of their perikaryons within the range of their dendrites.

It is well known that the axon of the lateral horn neurons form the preganglionic sympathetic fibers. However, it is not clear if the visceral afferents reach the lateral horn neurons directly or indirectly as yet. In the case of the latter possibility localization of the intercalated neurons should be determined.

Morphological and physiological studies (Hancock, Rigamonti and Bryan, 1973; Hancock, Foreman and Willis, 1975; Kerr, 1975) have shown that the cutaneous and visceral afferents terminate in many regions of the spinal cord. Yet, an exact source of the spinothalamic tracts is not clearly established as the tracts may take their origin from different laminae of the spinal cord.

It is well known that the lateral horn neurons receive input from the surpaspinal levels. Coote (1975) and Coote and Sato (1975) distinguished two neuronal arcs, viz., long and short ones. The long arc is that traveling through the brain stem whereas the short is arranged at the level of the spinal cord.

The aim of the present studies is to trace the range of dendrites of the neurons in the lateral horns within the 1st and 2nd lumbar segments. Basing on this we may conclude as to the input of the neurons.

Material and methods

The first and the second segments of the lumbar spinal cord were taken out from 19 cats (8 mature, 4 aged four days, 4 aged seven days and 3 cats twelve weeks of aged. The sections 100–250 µm thick were impregnated with the rapid Golgi method and according to Bubenaite's modification (Matsushita, 1969).

In the most cases the sections were taken in the transverse plane. In one mature cat the investigated segments were cut in the sagittal plane and in one cat they were cut in horizontal plane. In each group of age the lumbar segments were cut in horizontal plane. In each group of age the lumbar segments were cut into sections of 10–20 µm thick and stained with toluidine blue according to Nissl's method.

The results were based only on the GOLGI impregnated sections.

The graphic reconstructions were made with the aid of the microscope and photographs (GROTTEL, 1976, 1977).

Results

The neurons of the lateral horn present a uniform population of cells. They are distinctly separated from the lateral funiculus whereas medially they merge with the neurons of the intermediate gray. Certain sign in distinguishing this group is the appearance of small neurons. In the cases the axons of neurons were clearly impregnated the identification of lateral horn neurons could be established with certainly. These axons pass ventrally and somewhat laterally to the gray substance of the ventral horn (Figs. 1e and 2e).

Some neurons of the lateral horn located most laterally were found wihtin the lateral funiculus.

Considering the direction of dendrites of the lateral horn neurons some groups of them could be distinguished. The cranially and caudally directed dendrites are longer in the sagittal and horizontal planes than those in the transverse plane. The long axis of the cell body is usually parallel to the long axis of the spinal cord. In the sagittal and horizontal planes the arborizations of dendrites arising from the poles of neurons are observed.

In order to compare our investigations with the laminar pattern described by REXED (1952, 1954), descriptions of dendrites are based mainly on the transverse sections. In these sections the dorsally and ventrally running dendrites are close to the border between the dorsal and ventral horns and lateral funiculus as well (Figs. 1a, b, f and 2b, f). Dendrites passing dorsally enter often the gray substance of the dorsal horn in same cases reaching the lamina III (Figs. 1a and 2b). They branch always within the lateral half of the dorsal horn. Ventrally directed dendrites penetrate the ventral horn (Figs. 1h and 2h) reaching the lamina IX (Fig. 1g) and lamina VIII respectively.

Laterally oriented dendrites are always short and they pass into lateral funiculus (Figs. 1c, d and 2c, d). The dendrites passing medially vary as to their range and the course. They reach the lateral region of the intermediate gray (Figs. 1i, l and 2k, j), the central gray (Figs. 1m and 2l) as well as the thoracic

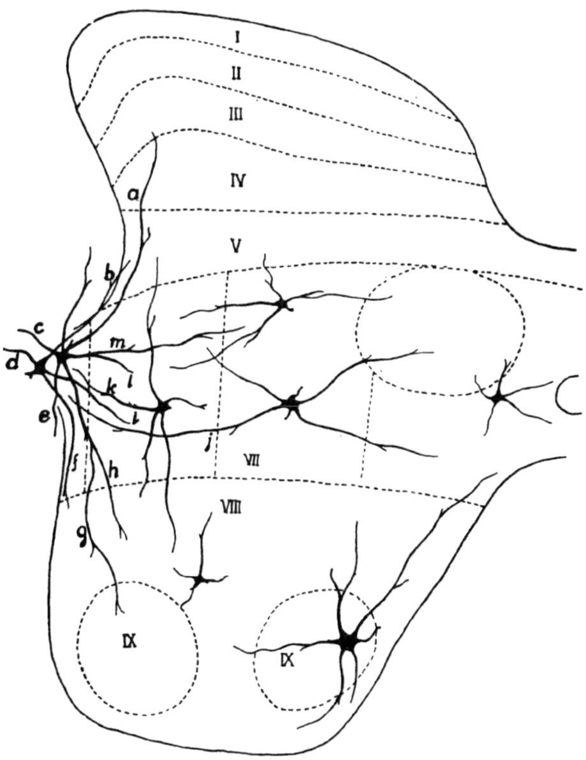

Fig. 1. Semischematic figure represents transverse section of right side of spinal cord at the level of L_1. Division into laminae is based on REXED's description (1952, 1954). In lamina VII (intermediate gray) paramedial, central and lateral parts have been distinguished. Cells were drawn from one section of 250 μm thick taken from a mature cat. Magn. × 80.

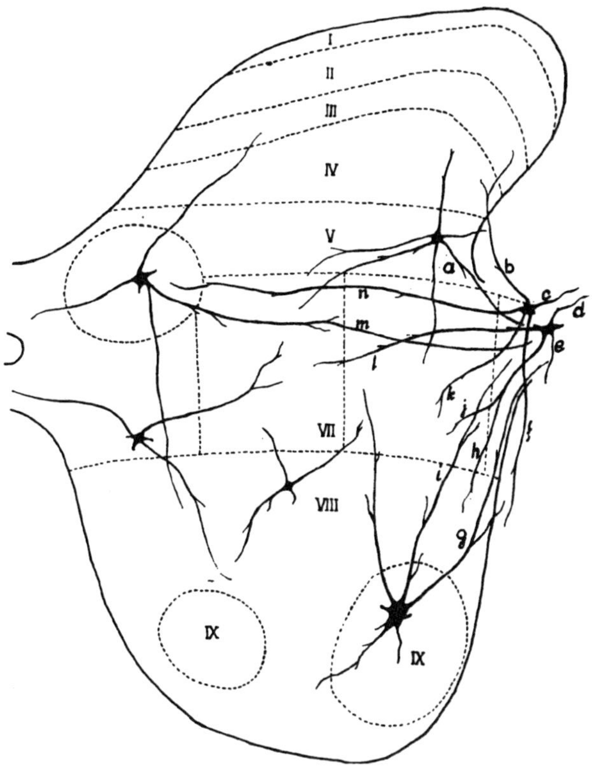

Fig. 2. Semischematic figure represents transverse section of left side of spinal cord at the level of L_1. Cells were drawn from one section of 250 μm thick taken from mature cat. Magn. × 80.

nucleus (Fig. 2n). Dendrites reaching the lateral region are most numerous (usually 2 to 3 in number), they are always short and directed rather ventrally. The longest dendrites passing medially reach the thoracic nucleus and those have not been found in the all sections. There were no dendrites directed medially with their range to the ventral part of the paramedial region.

The lateral horn receives many dendrites from different regions of the spinal gray. Dendrites from the dorsal horn take their origin in the lamina V (Fig. 2a). Those from ventral horn originate in the laminae VIII and IX (Fig 2g, i). The central (Fig. 1j) and lateral (Fig. 1k) regions of the intermediate gray as well as thoracic nucleus (Fig. 2m) send their dendrites to the lateral horn as well.

Discussion

According to RETHELYI (1972) neurons of the lateral horn are of 25×35 µm in size. In the present investigations the measurements of neurons were performed only in some cases. In such cases the results of Réthelyi were confirmed. Our results are also in accord with Réthelyi as to the arrangement of the longest dendrites of the neurons within the long axis of the spinal cord. The author also mentioned the transversely oriented dendrites. However, he did not describe their distribution.

The arrangement of dendrites within the transverse plane was the aim of the present studies. This is very important considering terminations of many neuronal tracts.

LARUELLE (1936) taking into consideration the shape and size of cells claimed that the autonomic centers posses the characteristic uniform population of cells.

Basing on the range of dendrites of cells it has to be pointed out that they differ markedly although in Nissl's stained sections the group of cells is distinctly separated especially from the lateral funiculus. Dendrites of each cell of the lateral horn were oriented in different directions. It was observed, however, that some of dendrites have similar course. It was shown in previous papers (GROTTEL, 1976, 1977) that dendrites of motonurons showed more uniform arrangement in comparisson to those of the lateral horn.

Considering our investigations and data from the literature it is very interesting to discuss afferents to the lateral horn. RÉTHELYI (1972) describes axons which pass transversely between the lateral funiculus and dorsal horn, and after they reach the lateral horn they turn along with the long axis of the spinal cord. He does not mention of synapses between axons and dendrites of the lateral horn neurons directed dorsally. The author traced the course of those axons in both directions with the light microscope. Using electron microscope he described the synapses of axons with dendrites and cell bodies of the lateral horn neurons only in their course along the axis of the spinal cord. One should be cautions to compile the light microscopic and electron microscopic results as to the course and connections of the nerve fibers. This unables the small area of the spinal cord available for the electron microscopic studies. In such a cases there is no certainly if the neuryte observed in the light microscope is the same one seen in the electron microscope. There is also possibility that longitudinally travelling axons described by RÉTHELYI are not the same observed by the author in the transverse sections of the spinal cord.

It is well known from studies of the motoneuron groups that they are related functionally in the long axis of the spinal cord. As a proof for the statement is the arrangement of the axons of Renshaw cells (JANKOWSKA and LINDSTRÖM, 1971) as well as these of interneurons which mediate Ia reciprocal inhibition of motoneurons (JANKOWSKA and LINDSTRÖM, 1972; JANKOWSKA and ROBERTS, 1972a, b). Axons of these cells pass into funiculi and affect the motoneurons in the large area in the long axis of the spinal cord.

It seems conceivable that the small cells situated medially to the lateral horn neurons may be considered as interneurons. This cannot be concluded, however, from the present studies for it requires neurophysiological investigations.

The course of dendrites presumes the direction of the afferent tract. It may be suggested then that the dorsally directed dendrites may receive inhibitory impulses from the ventrolateral reticular formation and from the caudal nucleus of raphe of the medulla. COOTE and MACLEOD (1975) using physiological methods have shown that axons of these nuclei, before the reached the lateral horn, ran in the dorsal part of the lateral funiculus in the homolateral half of the spinal cord. It is also probable that some of the ventrally directed dendrites receive also afferents from the ventromedial reticular formation the axons of which pass in the ventral part of the homolateral funiculus.

The descending tracts from the brain stem block the segmental somatovisceral reflexes. COOTE and SATO (1975) found that the intercostal reflex influencing the heart rate was inhibited by the descending tract from the ventromedial reticular nucleus of the medulla. The spinal reflexes are then blocked by the tonic activity of the brain stem centers. The same effect was elicited after section of the contralateral half of the spinal cord. In this case the authors transected also the spinothalmic tract. It is difficult to charge then that the reticülospinal tract is the only one responsible for the reflex. Hancock et al. (1973, 1975) have shown that the somatic and visceral afferents converged on the spinothalamic neurons. There is also evidence that the neurons providing origin to the long ascending tracts receive additional afferents from interneurons. This was described in details in the case of VSCT by Lindström (1973).

The damage of the spinothalamic tract may cause the appearance of the segmental visceral reflexes. The afferent tracts may inhibit higher inhibitory centers of the brain stem although the influence of different afferents is not the same. Lisander and Martner (1975) showed the facilitation of the intestinal movements after laparatomy. In this case the lack of somatic afferents may inhibit the sympathetic fibers reaching the intestinal tract.

Coote (1975) found that afferent fibers reaching the spinal cord are in 48—75 per cent unmyelinated. Réthelyi (1972) described that some of those fibers terminated in the intermediolateral nucleus. They entered the nucleus from the medial side. Our investigations showed that such fibers might join the medially directed dendrites. It is not known if the lateral horn neurons form synaptic connections with afferent fibers directly or by the way of interneurons. The latter way is more probable.

The lateral horn neurons receive many afferents from the dorsal radix, intraspinal tracts as well as from the supraspinal levels. Such connections are possible if we consider the wide range of dendrites of the cells which was shown in this studies. It is difficult, however, to presume which afferents synapse with particular dendrites.

The sympathetic system takes a main role in the regulation of the heart rate, blood distribution and respiratory activity during the muscle work. The afferents from the dorsal roots may be transmitted on dendrites of lateral horn neurons within the intermediate gray as well as in the thoracic nucleus or even in the dorsal horn. It is probable that through these ways pass the proprioceptive fibers into the lateral horn. Coote (1975) suggested that fibers transmitting impulses from joint receptors and muscular chemoreceptors reached the lateral horn neurons. Coote (1975) suggested that fibers transmitting impulses from joint receptors and muscular chemoreceptors reached the lateral horn neurons.

Basing on the degenerative studies of Nyberg-Hansen (1966) it is possible to claim that the dendrites of the lateral horn neurons may contact also with fibers of the corticospinal tract, contralateral rubrospinal tract (lateral parts of laminae V—VII), lateral vestibulospinal tract of the homolateral part of the spinal cord, bilateral medial vestibulospinal tracts as well as with bilateral interstitiospinal tracts (laminae VII and VIII).

It may be suggested that medially directed dendrites of the lateral horn neurons receive afferents which "omit" an exact source of the neurons giving origin to the spinothalamic and spinocerebellar tracts. In transmission of such information may take part some interneurons or offsets of ascending tracts of the spinal cord.

The above described suggestions were made on the basis of the arrangement of dendrites so on the morphological studies only. They require confirmation in the neurophysiological studies.

References

Coote, J. H.: Physiological significance of somatic afferent pathways from skeletal muscle and joints with reflex effects on the heart and circulation. Brain Res. **87**, 139—144 (1975).

Coote, J. H. and V. H. Macleod: The spinal route of sympatho-inhibitory pathways descending from the medulla oblongata. Pflügers Arch., **359**, 335—347 (1975).

Coote, J. H. and A. Sato: A role for a descending sympatho-inhibitory pathway on the ventral part of the spinal cord. J. Physiol., **252**, 21P—22P (1975).

Grottel, K.: Dendrite bundles of the ventral horns cells of the thoracic portion of the spinal cord in cats. Folia Morphol. (Warsz.), **35**, 393—403 (1976).

Grottel, K.: Distribution of dendrites in the ventral horn of the thoracic part of the spinal cord in the sagittal plane in cats. Folia Morph. (Warsz.), **36**, 29—36 (1977).

Grottel, K. and T. Hofman: Dendritic range of the neurons of the intermediate gray at the levels of the first and second lumbar segment of the spinal cord in the cat. 1. The range of dendrites of the central region neurons. — (in print).

Grottel, K. and T. Hofman: Dendritic range of the neurons of the intermediate gray at the levels of the first and second lumbar segment of the spinal cord in the cat. 2. The range of dendrites of the paramedial region neurons. — (in print).

Grottel, K. and T. Hofman: Dendritic range of the neurons of the intermediate gray at the levels of the first and second lumbar segments of the spinal cord in the cat. The range of dendrites of the lateral region neurons. — (in print).

Hancock, M. B., R. D. Foreman and W. D. Willis: Convergence of visceral and cutaneous input onto spinothalamic tract cells in the thoracic spinal cord of the cat. Exp. Neurol., **47**, 240—248 (1975).

Hancock, M. B., D. D. Rigamonti, and R. N. Bryan: Convergence in the lumbar spinal cord of pathways activated by splanchnic nerve and hind limb cutaneous nerve stimulation. Exp. Neurol. ,**38**, 337—348 (1973).

Jankowska, E. and S. Lindström: Morphological identification of Renshaw cells. Acta physiol. scand., **81**, 428—430 (1971).

JANKOWSKA, E. and S. LINDSTRÖM: Morphology of interneurones mediating Ia reciprocal inhibition of motoneurones in the spinal cord of the cat. J. Physiol., **226**, 805–823 (1972).

JANKOWSKA, E. and W. J. ROBERTS: An electrophysiological demonstration of the axonal projections of single spinal interneurones in the cat. J. Physiol., **222**, 597–622 (1972a).

JANKOWSKA, E. and W. J. ROBERTS: Synaptic actions of single interneurones mediating reciprocal Ia inhibition of motoneurones. J. Physiol. **222**, 623–642 (1972b).

KERR, F. W. L.: Neuroanatomical substrates of nociception in the spinal cord. Pain, **1**, 325–356 (1975).

LARUELLE, L. Contribution á l'etude du nevraxe végétatif. C. R. Ass. Anat. **31**, 210–229 (1936).

LINDSTRÖM, S.: Recurrent control from motor axon collaterals of Ia inhibitory pathways in the spinal cord of the cat. Acta physiol. scand., suppl. **392** (1973).

LISANDER, B. and J. MARTNER: Integrated somatomotor, cardiovascular and gastrointestinal adjustments induced from the cerebellar fastigial nucleus. Acta physiol. scand., **94**, 358–367 (1975).

MATSIUSHTA, M.: Some aspects of the interneuronal connec tions in cats spinal cord gray matter. Comp. Neurol., **136**, 57–80 (1969).

NYBERG-HANSEN, R.: Functional organization of descending supraspinal fibre systems to the spinal cord. Anatomical observations and physiological correlations. Reviews of Anatomy, Embryology and Cell Biology, Springer-Verlag, Berlin, Heidelberg, New York (1966).

RÉTHELYI, M.: Cell and neuropil architecture of the intermediolateral (sympathetic) nucleus of cat spinal cord. Brain Res., **46**, 203–213 (1972).

REXED, B.: The cytoarchitectonic organization of the spinal cord in the cat. J. Comp. Neurol., **96**, 415–466 (1952).

REXED, B.: A cytoarchitectonic atlas of the spinal cord in the cat. J. Comp. Neurol., **100**, 297–380 (1954).

The author's adress:

Dept. of Normal Anatomy Institute of Biostructure.
Academy of Medicine in Poznań,
60781 Poznań,
Av. Swięcickiego 6 Poland.

Buchbesprechung

Treatment of Neuromuscular

GRIGGS, R. C. and R. T. MOXLEY (Hrsg.): Treatment of Neuromuscular Diseases (Advances in Neurology, Band 17). Raven Press, New York 1977. 16 × 24 cm, XIV, 370 S., zahlr. Abb. u. Tab., Ganzl. $ 31,20. ISBN 0-89004-113-X

In 20 Einzelbeiträgen bemühen sich international als besonders kompetent ausgewiesene Autoren um eine „Up-to-date"-Darstellung der einzelnen Krankheitsgruppen im Hinblick auf gegenwärtige und zukünftige therapeutische Möglichkeiten. Die einleitenden Kapitel beinhalten auch die Fortschritte der Grundlagendisziplinen. Im Hauptteil werden die Therapie-Methoden der Wahl nach dem neuesten Stand und zum Schluß Forschungstrends mit Aussicht auf Ergebnisse, die der Behandlung nutzbar gemacht werden können, abgehandelt. Jedem Beitrag ist ein Literaturverzeichnis beigefügt. Am Ende des Buches findet sich ein Sachverzeichnis (6 Seiten).

Die klinische Untersuchungstechnik bei Patienten mit neuromuskulären Krankheiten wird hervorragend von BROOKE beschrieben. Auf funktionell-anatomischer Grundlage werden wertvolle Hinweise für einfache und zuverlässige Handgriffe und Provokationen im Rahmen der Inspektion, Palpation, Reflexauslösung und Prüfung der groben Kraft vermittelt, die z. T. in Lehrbüchern nicht zu finden sind.

A. S. PENN referiert die neueren Beiträge immunpathologischer Forschung zur Pathogenese der Myasthenia gravis pseudoparalytica, der Polymyositis und Dermatomyositis. In der Myastheniebehandlung stehen nach wie vor neben der Anwendung von Cholinesterasehemmern die medikamentöse Immunsuppression und die Thymektomie im Vordergrund. Letztere, deren Effektivität mit 75% angegeben wird, bewirkt eine bessere und zuverlässigere Ansprechbarkeit auf die medikamentöse Behandlung. Diagnostik und Therapie der Myositiden werden ausführlicher und speziell im folgenden Kapitel von L. P. ROWLAND, CH. CLARK und M. OLARTE abgehandelt. Die Autoren empfehlen die Prednisolontherapie mit einer Anfangsdosierung von 60 bis 100 mg täglich oder 100 mg alternierend jeden 2. Tag sowie in therapieresistenten Fällen die immunosuppressive Behandlung, vorzugsweise mit oraler Azathioprin-Anwendung in einer Dosierung von 150 mg täglich bei Erwachsenen. Schlußfolgerungen aus der Auswertung von 47 gezielt untersuchten Krankheitsverläufen bei Myasthenie veranlassen T. R. JOHNS in einem weiteren Beitrag, als Methode der Wahl die kombinierte Applikation von Kortikoiden und Thymektomie vorzuschlagen. Die Kortikoidtherapie ist nach Bedarf mit Cholinesterasehemmern zu kombinieren. Letztere allein wird nur bei leichten Verlaufsformen und guter Effektivität empfohlen.

Der aktuelle Erkenntnisstand über Myopathien, die auf Glykogen- und Fettstoffwechselstörungen beruhen, wird von S. DILLEURO und A. B. EASTWOOD dargestellt.

R. C. GRIGGS gibt einen kurzen, ausgezeichneten Überblick über die myotonischen Störungen und periodischen Lähmungen. Die wichtigsten Therapieempfehlungen lauten: Myotonia congenita: Quinin, Procainamid, Phenytoin. — Myotonische Dystrophie: keine wirksame Behandlung möglich. — Hypokaliämische Form der periodischen Lähmung: Acetazolamid, hyperkaliämische Form: Thiozid, Acetazolamid.

Die einer gezielten Therapie nicht zugängliche Gruppe der Muskeldystrophien ist Gegenstand mehrerer Kapitel, die Einzelergebnisse der Forschung, hypothetische Ansätze und interessante Anregungen enthalten (R. T. MOXLEY: Insulinstoffwechsel bei Muskeldystrophien, W. R. MARKESBERRY: Pathologie und virologische Übersicht, W. KING ENGEL detaillierte eigene Untersuchungsergebnisse und Interpretationen). Genetische Betrachtungen zu neuromuskulären Krankheiten werden von P. T. ROWLEY beigesteuert.

Im zweiten Teil des Buches werden die neurogenen Krankheitsformen besprochen. Unter den „therapierbaren" Neuropathien (M. J. BROWN) finden sich die in manchen Ländern häufigen Formen mit bekannter Ätiologie wie Lepra und Vitaminmangel, die einer kausalen Therapie zugänglich sind und die mannigfaltigen infektiös oder toxisch bedingten Krankheiten. Die Rolle der Kortikoidtherapie bei ätiologisch nicht geklärten bzw. kausaler Behandlung nicht zugänglichen Polyneuropathien wird differenziert und kritisch beurteilt. Auf chirurgische Indikationen wird kurz hingewiesen. Die sonstige symptomatische Behandlung wird in bekannter Weise erwähnt.

Die morphologischen Grundlagen als Voraussetzung zum Verständnis der peripheren Neuropathien werden von P. J. DYCK sehr anschaulich vermittelt. Verf. weist auf die besondere Bedeutung dreidimensionaler morphometrischer Studien hin.

Der klinische Neurologe wird das Kapitel „Behandlung der amyotrophischen Lateralsklerose" von D. GOLDBLATT mit besonderem Interesse lesen, in der Hoffnung, Fortschritte der Therapie dieser vielleicht grausamsten Krankheit zu erfahren. Leider ist das nicht der Fall, die Möglichkeiten der symptomatischen Behandlung und der Linderung des Leidens werden jedoch aus großer persönlicher Erfahrung heraus dargelegt.

Die abschließenden Beiträge runden die Thematik ab: eine kurze Darstellung der klinischen Elektromyographie (M. P. MC QUILLEN, eine ausführliche Diskussion zur infantilen Hypotonie („Floppy Baby", G. J. MYERS), Atemstörungen als Komplikation neuromuskulärer Erkrankungen (W. J. HALL), Kardiale Manifestationen neuromuskulärer Störungen (J. COHEN) und zur orthopädischen Korrektur von Deformierungen des Bewegungsapparates bei Muskeldystrophie (I. M. SIEGEL).

Den Herausgebern und Autoren ist es gelungen, den aktuellen Erkenntnisstand in bezug auf die neuromuskulären Krankheiten ausführlich genug aber straff, interessant und anschaulich, so darzustellen, daß der Kliniker das Buch als praktische Therapieempfehlung benutzen, sich aber auch über theoretische Zusammenhänge und Forschungstrends informieren kann und der Grundlagenforscher angeregt wird, offene Fragen aufzugreifen und Lücken zu schließen.

H. A. F. SCHULZE, Berlin

J. Hirnforsch. 19 (1978) 445–456

Aus dem Anatomischen Institut des Bereiches Medizin (Charité) der Humboldt-Universität Berlin
(Komissar. Direktor: Prof. Dr. sc. med. J. Staudt)

Zur umweltabhängigen Differenzierung von Pyramidenneuronen im Hippocampus (CA 1) der Ratte
Die Differenzierung von apikalen Seitendendriten und Basaldendriten

Von

Michael Frotscher, Kirsten Scharmacher und Michael Scharmacher

Mit 7 Abbildungen und 4 Tabellen

(Eingegangen am 20. Dezember 1977)

Zusammenfassung: Insgesamt 60 neugeborene Wistar-Ratten wurden in drei Gruppen aufgeteilt, von denen jede einer unterschiedlichen Umweltsituation ausgesetzt wurde. Die erste Gruppe wurde in völliger Dunkelheit gehalten, die zweite wurde einer Dauerbeschallung ausgesetzt, während die dritte Tiergruppe (Kontrollen) unter normalen Laborbedingungen aufgezogen wurde. Sämtliche Tiere wurdem am 15. Postnataltag getötet und nach der Golgi-Kopsch-Methode aufgearbeitet.

Apikale Seitendendriten und Basaldendriten von Pyramidenzellen der CA 1-Region des Hippocampus wurden für Dendritenlängenmessungen und Spineszählungen nach einem zentrifugalen Ordnungssystem klassifiziert.

Während die Dendritenlängenmessungen sowohl in einzelnen Dendritenordnungen als auch im Einzeldendritenfeld (sämtliche Verzweigungen eines vom apikalen Hauptdendriten oder vom Perikaryon entspringenden Dendriten) beim Gruppenvergleich keine statistisch signifikanten Unterschiede ergaben, wurden bei den Spineszählungen signifikante Veränderungen festgestellt. Im Vergleich mit den Kontrollen ist die Spinesanzahl bei den in Dunkelheit aufgewachsenen Tieren an apikalen Seitendendriten wie Basaldendriten signifikant vermindert, während beschallte Tiere im Kontrolltiervergleich signifikant mehr Spines an apikalen Seitendendriten und Basaldendriten aufweisen (mittlere Spinesanzahl im apikalen EDF: Kontrollen 30,1 $\pm$ 26,9, Dunkeltiere 17,1 $\pm$ 15,4 (−43,2%), beschallte Tiere 36,4 $\pm$ 31,4 (+20,9%); basales EDF: Kontrollen 82,2 $\pm$ 70,1, Dunkeltiere 47,3 $\pm$ 35,1 (−42,5%), beschallte Tiere 114,1 $\pm$ 94,6 (+38,8%)).

Die Ergebnisse bestätigen vorausgegangene Untersuchungen zur Spinesanzahl am apikalen Hauptdendriten von CA 1-Neuronen (Frotscher et al. 1975) und zur Synapsenanzahl im Stratum radiatum der CA 1-Region (Frotscher et al. 1977) an unter identischen experimentellen Bedingungen aufgezogenen Tieren. Nach diesen Ergebnissen beeinflussen die gegebenen Umweltbedingungen die Synaptogenese im Hippocampus, obgleich diese Region direkt weder in den optischen noch in den akustischen Analysator eingeschaltet ist.

Summary: A total of 60 newborn Wistar-rats was divided into three groups, each of which was exposed to a different environmental situation. The first group was reared in complete darkness while the second one was subjected to permanent noise. The third group (controls) was reared under normal laboratory conditions. The animals of all three groups were sacrificed on postnatal day 15 and prepared according to the Golgi-Kopsch technique.

Apical oblique dendrites and basal dendrites of hippocampal pyramidal cells (CA 1 region) were ranked in a centrifugal ordering system for measurements of dendritic length and for spine counts.

In the successive dendritic orders as well as in the dendritic field (all branches of a dendrite arising from the cell body or main apical shaft) no statistically significant differences between the groups were found with regard to dendritic length. The spine counts, however, revealed significant changes. In the dark reared group, the number of spines on oblique dendrites as well as on basal dendrites is significantly decreased in comparison with the controls. On the contrary, animals subjected to permanent noise exhibit significantly more spines on oblique and basal dendrites when compared with the controls (mean number of spines in the apical dendritic field: controls 30,1 $\pm$ 26,9, dark reared animals 17,1 $\pm$ 15,4 (−43,2%), animals subjected to noise 36,4 $\pm$ 31,4 (+20,9%); basal dendritic field: controls 82,2 $\pm$ 70,1, dark reared animals 47,3 $\pm$ 35,1 (−42,5%), animals exposed to noise 114,1 $\pm$ 94,6 (+38,8%)). These results confirm previous studies on the number of spines along the main apical shaft of CA 1 pyramidal cells (Frotscher et al. 1975) and on the number of synapses in the stratum radiatum of CA 1 (Frotscher et al. 1977) in rats reared under identical experimental conditions. On the basis of these results it seems to be obvious that synaptogenesis in the hippocampus is influenced by the environmental conditions used though this region is not directly involved in the optic or in the acoustic pathway.

1 Mit dankenswerter Unterstützung durch einen Forschungsauftrag des Ministeriums für das Hoch- und Fachschulwesen der DDR.

Seit den Untersuchungen von VALVERDE (1967, 1971) ist bekannt, daß Dunkelaufzucht in einer verzögerten Differenzierung der Spines an Neuronen des visuellen Cortex resultiert, Faktoren der äußeren Umwelt somit die Neuronogenese beeinflussen. Da Vergleichszählungen an Neuronen im Cortextemporalis keine von Kontrolltieren signifikant abweichenden Ergebnisse brachten (VALERDE 1967), wurde allgemein angenommen, daß es sich in diesen Experimenten um einen spezifischen auf den optischen Analysator beschränkten Effekt handelt. Inzwischen liegen im Schrifttum aber verschiedene Ergebnisse vor, die belegen, daß sensorische Deprivation oder Stimulation auch Neuronenpopulationen erfaßt, die nicht direkt dem entsprechenden Anaylsator angehören (COLEMAN and RIESEN 1968; VRENSEN and DE GROOT 1974, 1975; RYUGO et al. 1975; SCHÖNHEIT und SCHULZ 1977a, b; FROTSCHER et al. 1975, 1977; FRIMMEL 1977). Demnach ist — zumindest morphologisch — nachgewiesen, daß Umweltreize in sehr komplexer Form die Differenzierung zentralnervöser Strukturen beeinflussen.

In der vorliegenden Arbeit wird mit der quantitativen Anaylse von Neuronen des Hippocampus nach Dunkelaufzucht (sensorische Deprivation) und Dauerbeschallung (Stimulation) gleichfalls eine Region untersucht, die keine direkten Verbindungen zu den beanspruchten Analysatoren besitzt. Es sollte geprüft werden, ob diese Umweltmodifikationen die postnatale Differenzierung der Pyramidenzellen des Hippocampus, der für die komplexe Verarbeitung sensorischer Afferenzen von Bedeutung ist (GREEN 1964; ADEY 1967; MEISSNER 1966; SMYTHIES 1967), beeinflussen.

In einer vorausgegangenen ersten Untersuchung wurde eine signifikante Verminderung der dendritischen Spines an apikalen Hauptdendriten von Pyramidenzellen der CA 1-Region bei Versuchstieren festgestellt, die vom 1.—15. Lebenstag in Dunkelheit gehalten worden waren. Vom 1.—15. Lebenstag dauerbeschallte Tiere wiesen dagegen eine signifikante Vermehrung der Spines im Kontrolltiervergleich auf (FROTSCHER et al. 1975). In parallel dazu durchgeführten elektronenmikroskopischen Untersuchungen wurde bei den Dunkeltieren eine signifikante Verminderung, bei den beschallten Tieren eine signifikante Erhöhung der Synapsenanzahl im Stratum radiatum der CA 1-Region gefunden (FROTSCHER et al. 1977). Da für die elektronenmikroskopischen Untersuchungen Aufnahmen verwendet wurden, die aus den zwischen den Hauptdendriten gelegenen Neuropilabschnitten stammen, wurde angenommen, daß sich die in den GOLGI-Untersuchungen gefundenen Veränderungen am Hauptdendriten auch auf die Seitendendriten der CA 1-Neurone erstrecken.

Mit den vorliegenden quantiattiven Untersuchungen an apikalen Seitendendriten und Basaldendriten von CA 1-Pyramidenneuronen werden die vorausgegangenen licht- und elektronenmikroskopischen Experimente überprüft und weitergeführt.

Material und Methodik[1]

Die Untersuchungen wurden an insgesamt 60 neugeborenen Wistar-Ratten durchgeführt, die in drei Versuchstiergruppen aufgeteilt wurden. Die Tiere waren in gleichartigen Boxen untergebracht und erhielten Standardfutter sowie Trinkwasser ad libitum. Zur Klärung des Einflusses unterschiedlicher Umweltreize auf die postnatale Differenzierung hippocampaler Pyramidenneurone wurden die Tiere der ersten Gruppe von Geburt an in völliger Dunkelheit gehalten, die Tiere der zweiten Gruppe einer Dauerbeschallung (ca. 80 db) ausgesetzt, während die Tiere der dritten Gruppe als Kontrollen unter normalen Laborbedingungen aufgezogen wurden. (vgl. auch FROTSCHER et al. 1975, 1977). Sämtliche Tiere wurden am 15. Lebenstag durch transcardiale Perfusionsfixierung in Äthernarkose getötet und nach einer modifizierten Golgi-Kopsch-Methode aufgearbeitet. Nach Imprägnation und Einbettung in Celloidin wurden Frontalschnittserien von 150 µm Schnittdicke angefertigt. Für die Auswertung wurden nach Zufallskriterien pro Tiergruppe 5 Gehirne ausgewählt. Von jeder Tiergruppe wurden jeweils 15 vollständig imprägnierte CA 1-Neurone (3 Zellen pro Gehirn) mittels Zeicheneinrichtung nach Abbé bei 600facher Vergrößerung gezeichnet (vgl. Abb. 1). Am Präparat wurden bei 640facher Vergrößerung die mikroskopisch sichtbaren Spines an Basaldendriten, apikalen Seitendendriten und den Dendriten des apikalen Büschels (Endaufzweigung des Hauptdendriten) gezählt. An Hand der Zeichnungen wurden die Dendriten nach Ordnungen entsprechend ihrem Verzweigungsgrad eingeteilt und mit einem Kurvimeter ihre Längen gemessen (vgl. ENGLISCH et al. 1974). Die statistische Auswertung umfaßte für jede Tiergruppe die Bestimmung der mittleren Spinesanzahl, der mittleren Spinesdichte (Spines/µm Dendritenlänge) und der mittleren Dendritenlänge von Basaldendriten, apikalen Seitendendriten und von Dendriten des apikalen Büschels in den einzelnen Ordnungen sowie in den Einzeldendritenfeldern (unter einem Einzeldendritenfeld (EDF) wird die Summe aller von einem Dendritenstamm abgehenden Aufzweigungen verstanden).

Zur Prüfung auf Homogenität innerhalb der Gruppen wurde die einfache Varianzanalyse angewendet; der Vergleich zwischen den Gruppen erfolgte mit dem t-Test nach STUDENT.

Ergebnisse

In der vorliegenden Arbeit wurden für die quantitative Analyse der Dendritenlängen und des Spinesbesatzes die Dendriten entsprechend ihrer fortschreitenden Verzweigung in einzelne Ordnungen

1 Frau Brigitte MANNSFELD danken wir für die sorgfältige technische Mitarbeit.

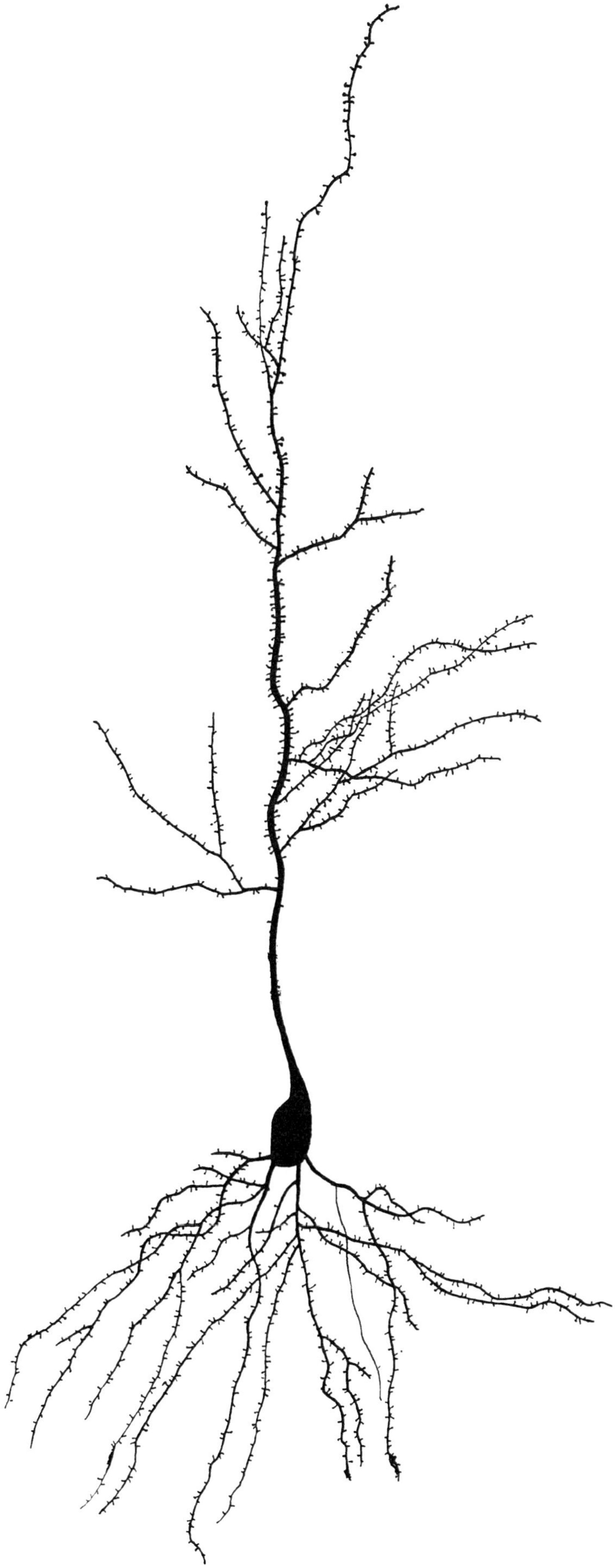

Abb. 1. Pyramidenzelle aus der CA 1-Region des Hippocampus einer 15 Tage alten Ratte (Kontrolltier). Nach dem Mikroskop bei 600facher Vergrößerung gezeichnet und photographisch verkleinert.

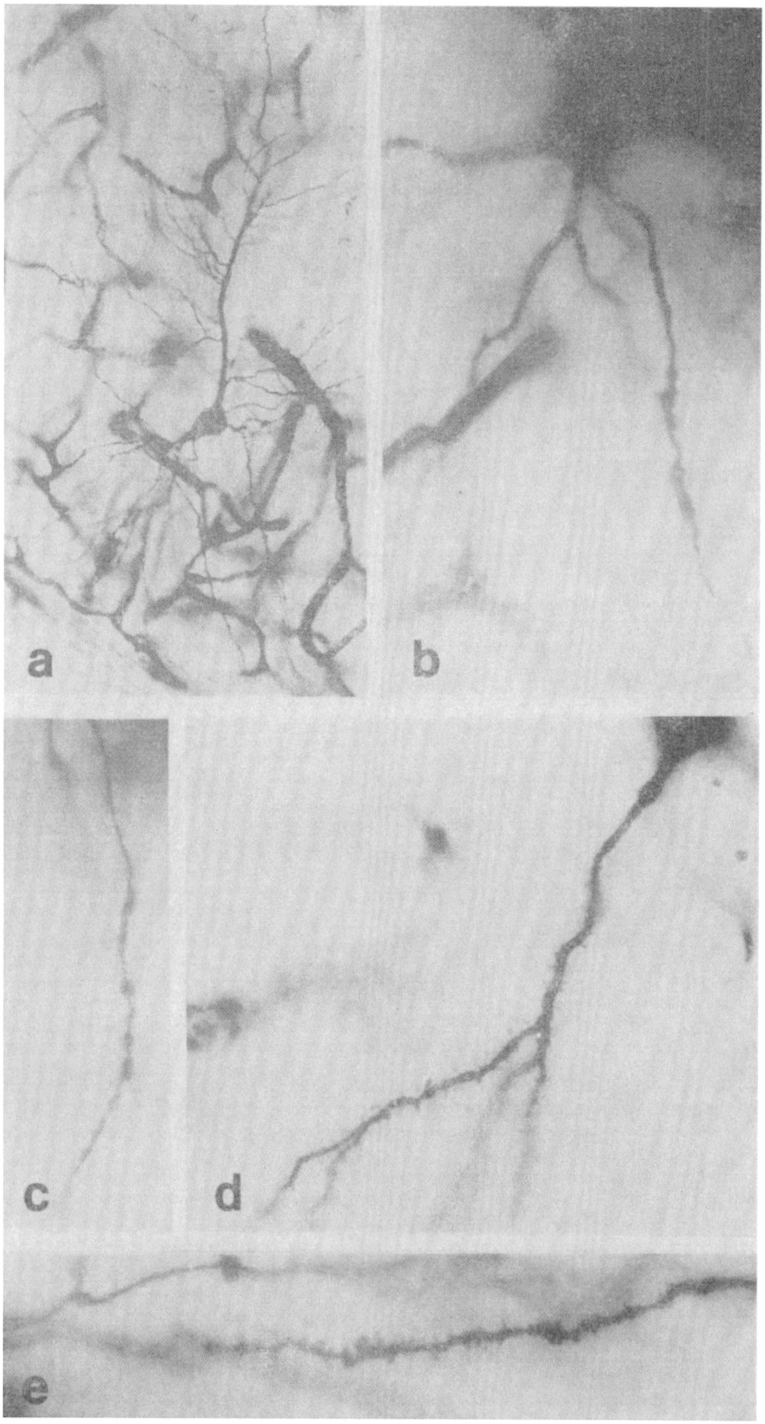

Abb. 2. a) Übersichtsaufnahme eines gesamten Pyramidenneurons aus der CA 1-Region bei einer 15 Tage alten Ratte (Kontrolltier). Golgi-Kopsch, Schnittdicke 150 µm. Vergrößerung etwa 190fach. b—e) Basaldendriten von CA 1-Neuronen. Vergr. 960fach.
b und c) 15 Tage altes, von Geburt an in Dunkelheit aufgezogenes Tier. Im Verlauf der Dendriten noch wachstumsbedingte Fortsatzanschwellungen sichtbar. Kaum Spines.
d) 15 Tage altes beschalltes Tier.
e) 15 Tage altes Kontrolltier. Bei beschalltem Tier wie bei Kontrolltier zahlreiche Spines, keine wachstumsbedingten Anschwellungen mehr.

aufgeteilt, wobei die primäre Abzweigung vom Hauptdendriten oder vom Perikaryon der 1. Ordnung entspricht (vgl. Englisch et al. 1974). Andererseits wurde die Gesamtheit aller Ramifikationen eines vom Hauptdendriten bzw. Perikaryon abzweigenden Dendritenstammes zum Einzeldendritenfeld zusammengefaßt. Die Einteilung der Dendriten in einzelne Ordnungen ist unseres Erachtens insbesondere bei Vergleichen unterschiedlicher experimenteller Gruppen günstig, da hiermit mögliche Unterschiede zwischen den Gruppen auf bestimmte Abschnitte des Dendritenbaumes lokalisiert werden können. Die Zusammenfassung zu Einzeldendritenfeldern trägt der funktionellen Bedeutung dieser Einheiten Rechnung (Aršavskij et al. 1971; Bogoslovskaja et al. 1971). Mit diesem Vorgehen wird an vorausgegangene quantitative Untersuchungen zur Struktur hippocampaler Neurone adulter Ratten (Englisch et al. 1974) und an quantitative Untersuchungen zur Ontogenese hippocampaler Neurone (Minkwitz 1976) angeknüpft. Die Ergebnisse zur experimentellen Differenzierungsbeeinflussung hippocampaler Neurone der vorliegenden Arbeit können daher mit Daten der normalen ontogenetischen Entwicklung dieser Zellen (MINKWITZ 1976) verglichen werden.

1. Dendritenmessungen

Die Werte für die mittlere Gesamtdendritenlänge im apikalen und basalen EDF bei den 15 Tage alten Normaltieren (Kontrollen) der vorliegenden Untersuchung ($90{,}7 \pm 78{,}2$ µm für das apikale, $321{,}3 \pm 229{,}7$ µm für das basale EDF, vgl. Tab. 4) lassen sich gut zwischen die von MINKWITZ (1976) für den 10. und 20. postnatalen Tag ermittelten Daten einordnen (10. Tag: apikales EDF $54{,}6 \pm 34{,}3$ µm, basales EDF $219{,}6 \pm 141{,}3$ µm; 20. Tag: apikales EDF $148{,}1 \pm 86{,}9$ µm, basales EDF $520{,}0 \pm 227{,}2$ µm).

Der statistische Vergleich der drei Versuchsgruppen der vorliegenden Untersuchung ergab bei den Längen der Basaldendriten, der apikalen Seitendendriten sowie bei den Dendriten des apikalen Büschels — sowohl in den einzelnen Dendritenordnungen als auch in den Einzeldendritenfeldern — keine signifikanten Unterschiede zwischen Kontrolltieren und Dunkeltieren einerseits und Kontrolltieren und beschallten Tieren andererseits. Bei subjektiver Betrachtung fällt auf, daß im Gegensatz zu den beiden anderen Tiergruppen Dendriten der Dunkeltiere (insbesondere Basaldendriten) häufig variköse Anschwellungen aufweisen, wie sie für frühe Stadien wachsender Neurone (CAJAL 1906, STENSAAS 1967, MOREST 1969, FROTSCHER 1975) beschrieben wurden (vgl. Abb. 2). Dieser Befund erhält eine gewisse Unterstützung dadurch, daß bei den Dunkeltieren die mittleren Dendritenlängen in den einzelnen Ordnungen wie auch im EDF (Basaldendriten!) die Werte der beiden anderen Gruppen wiederholt nicht erreichen (vgl. Tab. 2 und 4), so daß sich ein Differenzierungsrückstand der Dendriten der Dunkeltiere andeutet.

Zur Einschätzung des *Verzweigungsgrades* wurde die Anzahl der freien Dendritenendigungen/Dendriten 1. Ordnung (Verzweigungsindex, EAYRS 1955) ermittelt. Auch bei der Anwendung dieses morphologischen Parameters konnten keine statistisch signifikanten Unterschiede zwischen den Versuchsgruppen festgestellt werden. Die Ergebnisse zeigen aber den prinzipiell höheren Verzweigungsgrad der Basaldendriten auf (Verzweigungsindex apikaler Seitendendriten: Kontrollen $1{,}7 \pm 0{,}6$, Dunkeltiere $1{,}9 \pm 0{,}7$, beschallte Tiere $1{,}6 \pm 0{,}3$; Basaldendriten: Kontrolltiere $5{,}9 \pm 2{,}8$, Dunkeltiere $6{,}2 \pm 2{,}7$, beschallte Tiere $5{,}5 \pm 3{,}4$).

Mit der Bestimmung der Häufigkeit einzelner Dendritenordnungen werden ebenfalls Angaben über das dendritische Verzweigungsmuster vermittelt (Abb. 3). Unterschiede zwischen den Gruppen fehlen auch hier, wie bereits aus dem Vergleich der Dendritenlängen in den einzelnen Ordnungen und in den Einzeldendritenfeldern zu erwarten ist. Dagegen lassen sich die beschriebenen Unterschiede zwischen apikalen Seitendendriten und Basaldendriten nachweisen (ENGLISCH et al. 1974; MINKWITZ 1976). Bei den apikalen Seitendendriten liegt der Häufigkeitsgipfel bei den Dendriten 1. Ordnung und fällt dann ab. Hierbei ist zu berücksichtigen, daß nach MINKWITZ (1976) eine relativ große Anzahl unverzweigter apikaler Seitendendriten vorkommt. Bei den Basaldendriten findet man einen Anstieg bis zur 3. Ordnung und dann erst einen Abfall. Dieser Verzweigungsmodus scheint ein relativ invariantes Merkmal zu sein, da neben hippocampalen Pyramidenzellen (ENGLISCH et al. 1974; MINKWITZ 1976) auch Pyramidenzellen des visuellen bzw. sensorischen Cortex (WENZEL et al. 1973; KUNZ et al. 1974; UYLINGS 1977) und der vorderen cingulären Rinde (MICHALSKI et al. 1976) diese für apikale Seitendendriten und Basaldendriten unterschiedliche Häufigkeitsgipfel aufweisen.

Tab. 1. Spinesanzahl, Spinesdichte und Dendritenlänge in den einzelnen Ordnungen apikaler Seitendendriten von CA1-Neuronen bei Kontrolltieren, Dunkeltieren und beschallten Tieren (arithmetische Mittelwerte und Standardabweichungen)
S Spinesanzahl, SD Spinesdichte, DL Dendritenlänge
$^{+}$ signifikanter Unterschied zum Kontrolltierwert bei $p < 0{,}05$
$^{++}$ signifikanter Unterschied zum Kontrolltierwert bei $p < 0{,}01$
$^{-}$ kein signifikanter Unterschied zum Kontrolltierwert

	Kontrolltiere			Dunkeltiere			beschallte Tiere		
Ordnung	S	SD	DL	S	SD	DL	S	SD	DL
1	$14{,}4 \pm 12{,}7$	0,38	$41{,}1 \pm 31{,}0$	$6{,}6 \pm 4{,}6^{++}$	$0{,}22^{++}$	$35{,}2 \pm 27{,}5^{-}$	$18{,}9 \pm 12{,}9^{++}$	$0{,}44^{+}$	$50{,}8 \pm 33{,}6^{++}$
2	$12{,}7 \pm 9{,}7$	0,36	$40{,}9 \pm 27{,}6$	$6{,}4 \pm 4{,}7^{++}$	$0{,}23^{++}$	$33{,}1 \pm 21{,}9^{++}$	$15{,}5 \pm 9{,}9^{+}$	$0{,}44^{++}$	$41{,}1 \pm 26{,}6^{-}$
3	$10{,}8 \pm 11{,}2$	0,34	$34{,}7 \pm 25{,}5$	$7{,}0 \pm 4{,}6^{++}$	$0{,}20^{-}$	$37{,}4 \pm 21{,}0^{-}$	$18{,}0 \pm 10{,}0^{++}$	$0{,}44^{++}$	$42{,}9 \pm 20{,}0^{-}$
4	$7{,}9 \pm 5{,}3$	0,29	$29{,}7 \pm 20{,}6$	$6{,}3 \pm 3{,}8^{-}$	$0{,}22^{-}$	$32{,}8 \pm 20{,}9^{-}$	$18{,}1 \pm 8{,}7^{++}$	$0{,}44^{+}$	$41{,}4 \pm 16{,}3^{-}$
5	$5{,}7 \pm 6{,}3$	0,25	$20{,}5 \pm 22{,}8$	$8{,}4 \pm 2{,}1^{-}$	$0{,}27^{-}$	$32{,}0 \pm 5{,}8^{-}$	$15{,}5 \pm 6{,}8^{+}$	$0{,}63^{-}$	$30{,}0 \pm 8{,}8^{-}$

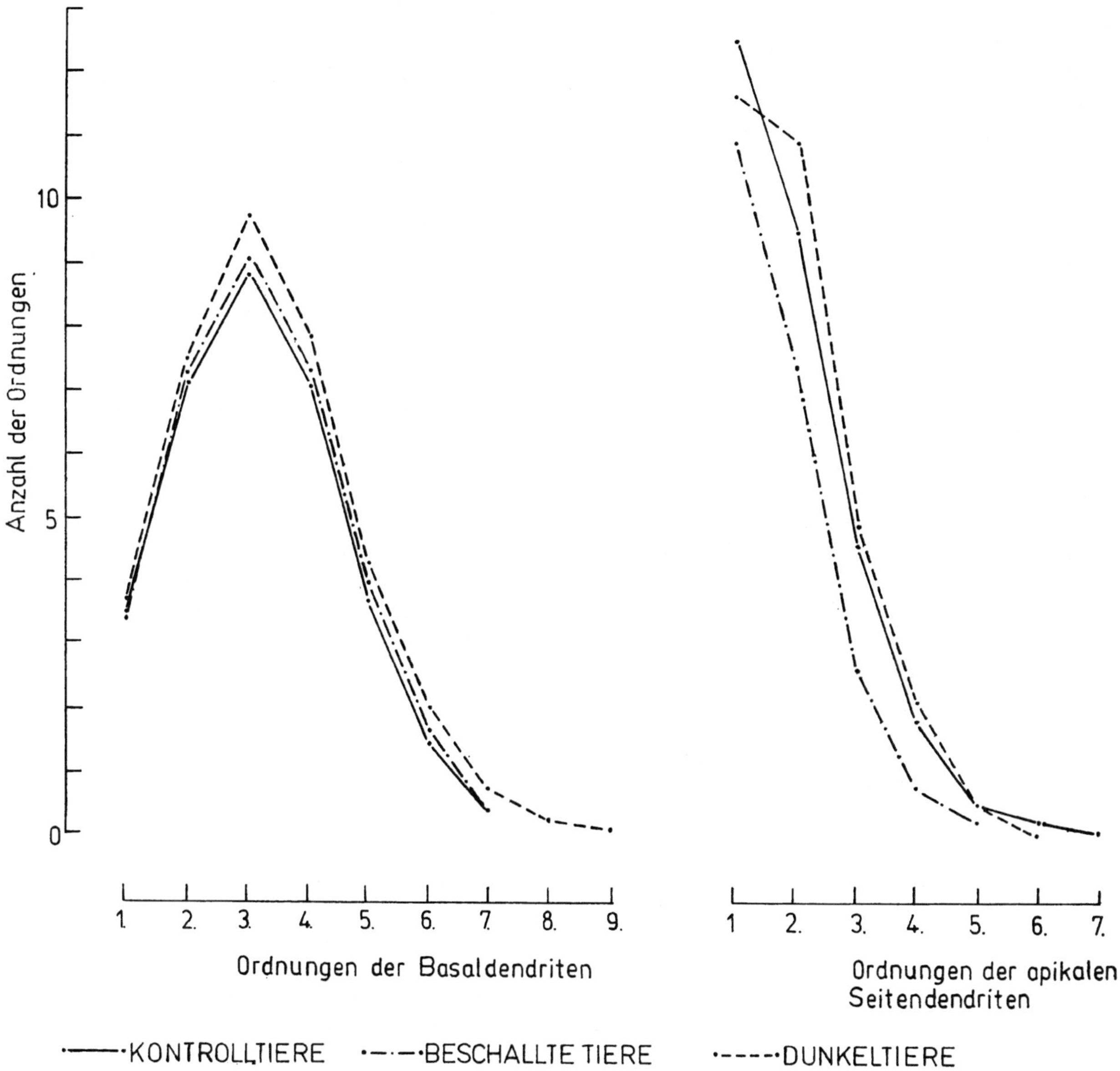

Abb. 3. Graphische Darstellung der mittleren Anzahl aufeinanderfolgender Ordnungen (Basaldendriten, apikale Seitendendriten) pro CA 1-Neuron bei Kontrolltieren, beschallten Tieren und Dunkeltieren.

2. *Spineszählungen*

Der statistische Vergleich der Spineswerte an Basaldendriten und apikalen Seitendendriten ergab signifikante Unterschiede zwischen Kontrolltieren und Dunkeltieren sowie zwischen Kontrolltieren und beschallten Tieren. Demnach kommt es unter den gegebenen Umweltbedingungen zu Abweichungen in der Spinesdifferenzierung. Im Vergleich mit den Kontrollen wurden bei den Dunkeltieren signifikant weniger Spines gezählt; bei den beschallten Tieren wurden dagegen signifikant höhere Spineswerte ermittelt (Abb. 2, 4—7). Die Veränderungen treten dabei jeweils gleichsinnig an apikalen Seitendendriten, Basaldendriten und an Dendriten des apikalen Büschels auf und erfassen die verschiedenen Ordnungen. In den Tabellen 1—3 sind die entsprechenden Mittelwerte für die einzelnen Dendritenordnungen sowie die Signifikanzen und die Umrechnung auf Spines/μm Dendritenlänge (Spines—Dendriten—Quotient oder Spinesdichte) angegeben (insbesondere am apikalen Büschel treten signifikante Unterschiede zwischen den Gruppen wiederholt nur im Spines—Dendriten—Quotienten auf, vgl. Tab. 3).

Obgleich sich die Unterschiede nicht in allen Dendritenordnungen als signifikant nachweisen lassen, kann unter Einbeziehung der Ergebnisse am apikalen Hauptdendriten (Frotscher et al. 1975) geschlußfolgert werden, daß die gegebenen Umweltmodifi-

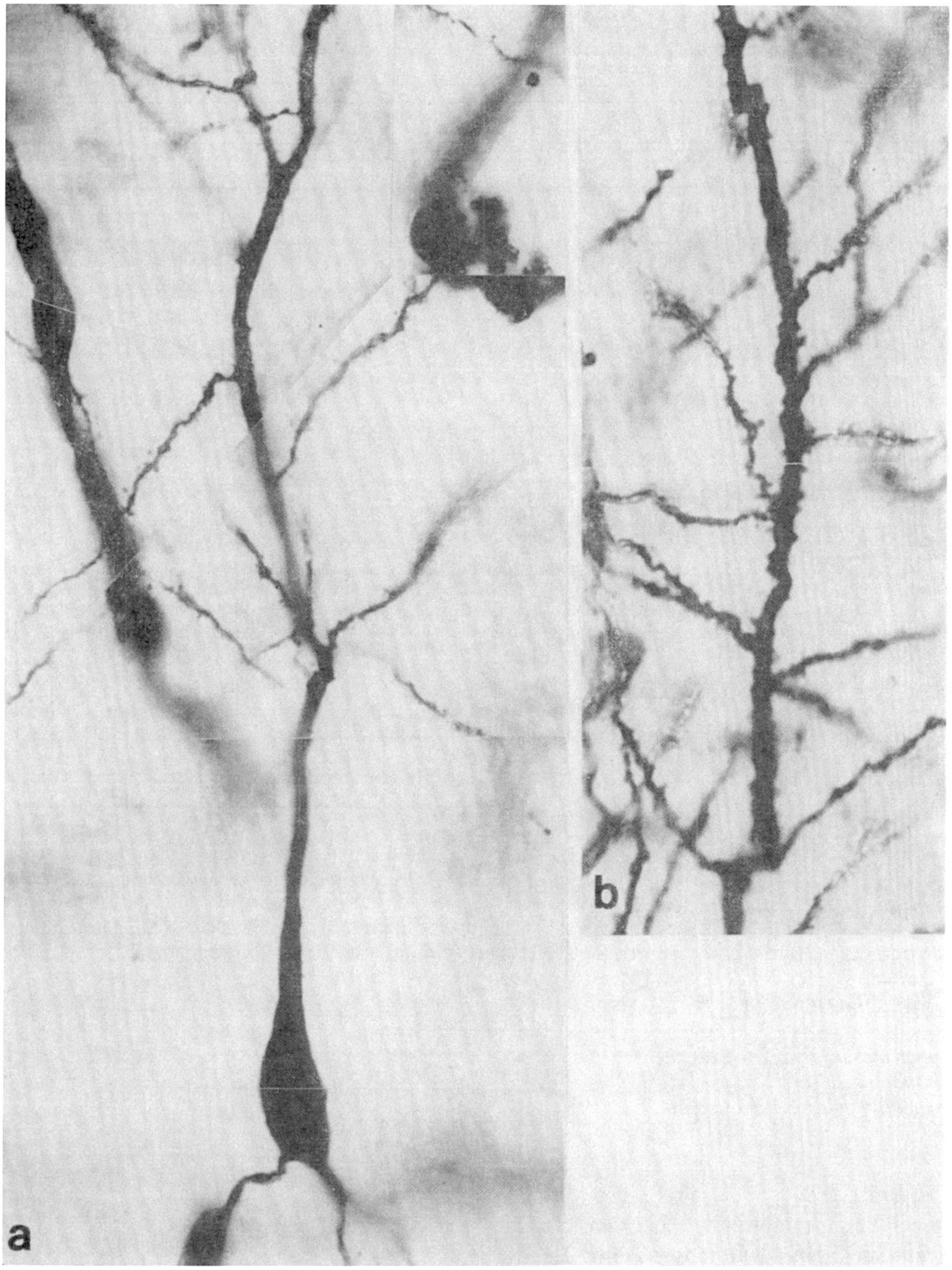

Abb. 4. Apikale Hauptdendriten und Seitendendriten von CA 1-Neuronen. Verg. 960fach.
a) in Dunkelheit aufgezogenes Tier
b) beschalltes Tier
Geringer Spinesbesatz, insbesondere am Hauptdendriten bei dem in Dunkelheit aufgezogenen Tier.

kationen die Differenzierung des gesamten CA 1-Neurons beeinflussen. Entsprechend sind die Veränderungen besonders ausgeprägt in den Einzeldendritenfeldern (Abb. 7, Tab. 4). Die mittlere Anzahl der Spines pro apikales EDF ist bei den Dunkeltieren um 43,2% signifikant ($p < 0,01$) niedriger als bei den Kontrollen. Bei den beschallten Tieren dagegen wurde eine signifikante Erhöhung (+20,9%) der Spinesanzahl im apikalen EDF vorgefunden ($p < 0,05$). Ähnlich liegen die Verhältnisse bei den basalen Einzeldendritenfeldern. Die Spinesanzahl bei den Dunkeltieren ist um 42,5% signifikant vermindert ($p < 0,01$) gegenüber einer Spinesvermehrung um 38,8% bei den beschallten Tieren ($p < 0,05$). Im EDF des apikalen Büschels traten bei den Dunkel-

Tab. 2. Spinesanzahl, Spinesdichte und Dendritenlänge in den einzelnen Ordnungen der Basaldendriten von CA 1-Neuronen bei Kontrolltieren, Dunkeltieren und beschallten Tieren (arithmetische Mittelwerte und Standardabweichungen) Abkürzungen s. Tab. 1

	Kontrolltiere			Dunkeltiere			Beschallte Tiere		
Ordnung	S	SD	DL	S	SD	DL	S	SD	DL
1	$2,7 \pm 5,0$	0,11	$19,6 \pm 19,1$	$1,4 \pm 2,7^{-}$	$0,08^{+}$	$16,3 \pm 16,2^{-}$	$3,0 \pm 6,4^{-}$	$0,12^{-}$	$16,2 \pm 20,5^{-}$
2	$8,6 \pm 9,2$	0,24	$32,6 \pm 30,0$	$4,7 \pm 4,5^{++}$	$0,17^{++}$	$28,5 \pm 23,7^{-}$	$9,5 \pm 9,7^{-}$	$0,31^{++}$	$32,9 \pm 29,9^{-}$
3	$11,3 \pm 11,1$	0,3	$40,9 \pm 31,8$	$6,4 \pm 4,6^{++}$	$0,21^{++}$	$35,4 \pm 28,5^{-}$	$12,8 \pm 9,8^{-}$	$0,39^{++}$	$36,6 \pm 26,2^{-}$
4	$10,2 \pm 9,7$	0,25	$42,6 \pm 31,5$	$5,3 \pm 4,0^{++}$	$0,19^{++}$	$31,2 \pm 24,0^{++}$	$14,8 \pm 10,8^{++}$	$0,38^{++}$	$42,8 \pm 27,1^{-}$
5	$9,8 \pm 6,8$	0,34	$34,7 \pm 24,2$	$6,1 \pm 4,7^{++}$	$0,20^{++}$	$36,3 \pm 28,4^{-}$	$13,6 \pm 9,5^{+}$	$0,37^{-}$	$42,9 \pm 27,1^{-}$
6	$10,1 \pm 6,9$	0,45	$43,5 \pm 22,6$	$4,7 \pm 4,6^{++}$	$0,19^{++}$	$23,6 \pm 20,1^{++}$	$17,4 \pm 10,3^{++}$	$0,39^{-}$	$34,5 \pm 26,3^{-}$

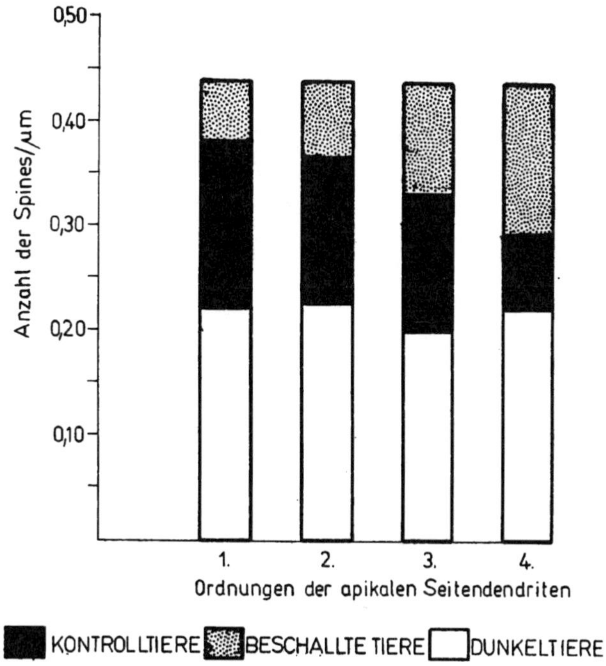

Abb. 5. Mittlere Spinesdichte an apikalen Seitendendriten aufeinanderfolgender Ordnungen (die Mittelwerte für die 3 Tiergruppen sind jeweils übereinander dargestellt).

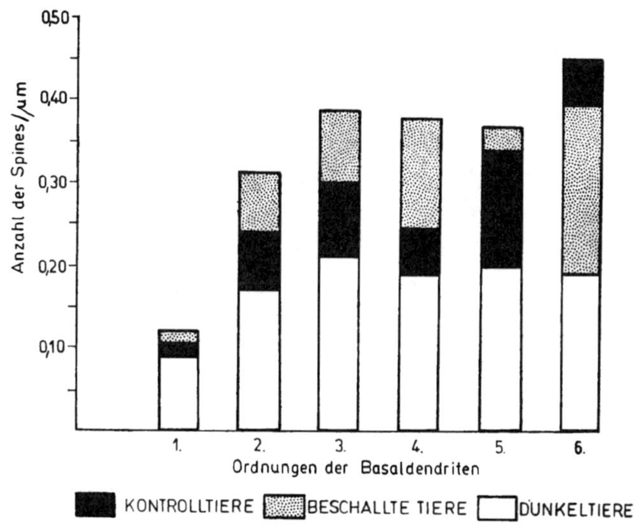

Abb. 6. Mittlere Spinesdichte an Basaldendriten aufeinanderfolgender Ordnungen bei den 3 Tiergruppen.

tieren signifikante Unterschiede sowohl in Spinesanzahl als auch Spinesdichte auf, während bei den beschallten Tieren nur die Spinesdichte signifikant verändert ist (0,43 Spines/µm gegenüber 0,34 Spines/µm bei den Kontrollen ($p < 0,05$)). Hierbei ist jedoch zu berücksichtigen, daß die Auswertung des apikalen Büschels wegen stärkerer Faserüberlagerung im Stratum lacunosum — moleculare häufig — insbesondere in den höheren Ordnungen — Schwierigkeiten bereitete (statistisch verwertbare Daten liegen daher nur bis zur 3. Dendritenordnung vor).

Ähnlich wie die Dendritenlängenmessungen lassen sich auch die Ergebnisse der Spineszählungen der vorliegenden Untersuchung gut mit den Daten der normalen Ontogenese (Minkwitz 1976) vergleichen. Während für das apikale EDF bei 10 Tage alten Rat-

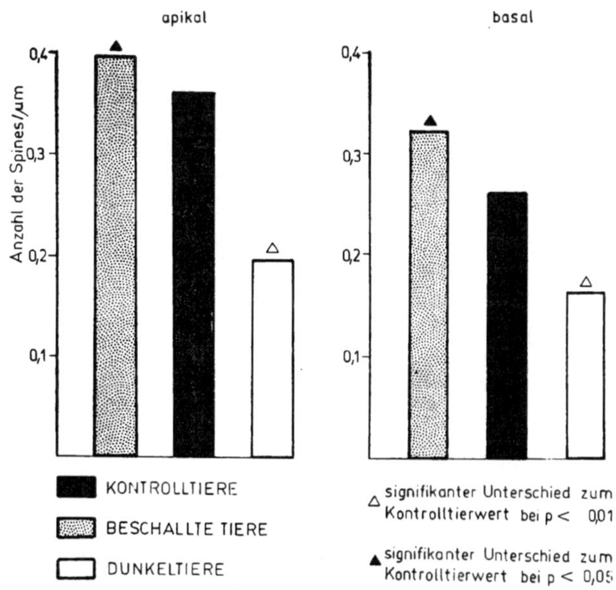

Abb. 7. Mittlere Spinesdichte im apikalen und basalen Einzeldendritenfeld der CA 1-Neurone bei Kontrolltieren, beschallten Tieren und Dunkeltieren.

Tab. 3. Spinesanzahl ,Spinesdichte und Dendritenlänge in den einzelnen Ordnungen des apikalen Büschels von CA 1-Neuronen bei Kontrollen, Dunkeltieren und beschallten Tieren (arithmetische Mittelwerte und Standardabweichungen) Abkürzungen s. Tab. 1

	Kontrolltiere			Dunkeltiere			beschallte Tiere		
Ordnung	S	SD	DL	S	SD	DL	S	SD	DL
1	$18,0 \pm 16,1$	0,41	$55,0 \pm 54,4$	$8,9 \pm 6,7^{+}$	$0,21^{++}$	$50,9 \pm 40,5^{-}$	$15,2 \pm 11,1^{-}$	$0,52^{-}$	$34,0 \pm 24,9^{-}$
2	$13,8 \pm 11,5$	0,31	$47,2 \pm 39,6$	$8,7 \pm 5,9^{-}$	$0,21^{++}$	$46,1 \pm 33,7^{-}$	$14,4 \pm 9,1^{-}$	$0,45^{+}$	$42,6 \pm 32,4^{-}$
3	$9,4 \pm 6,4$	0,32	$36,4 \pm 26,0$	$7,5 \pm 4,6^{-}$	$0,20^{++}$	$41,7 \pm 25,4^{-}$	$16,9 \pm 11,3^{++}$	$0,45^{+}$	$38,3 \pm 22,0^{-}$

ten ein Mittelwert von $9,6 \pm 7,9$ Spines ermittelt wurde, liegt der Wert bei 20 Tage alten Tieren bereits bei $83,7 \pm 52,2$ Spines. Der Wert von $30,1 \pm 26,9$ Spines für die 15 Tage alten Kontrolltiere der vorliegenden Untersuchungen läßt sich somit gut einordnen. Adulte, 3 Monate alte Ratten haben im Mittel $92,4 \pm 72,1$ Spines im apikalen EDF (Englisch et al. (1974). Demnach kommt es auch noch nach dem 20. Tag zu einer — wenn auch geringeren — Zunahme der Spines an apikalen Seitendendriten. Ähnlich liegen die Verhältnisse im basalen EDF (10 Tage alte Tiere $40,4 \pm 27,3$ Spines, 15 Tage alte Ratten $82,2 \pm 70,1$ Spines, 20 Tage alte Ratten $293,5 \pm 127,1$ Spines und $318,9 \pm 132,9$ Spines für 3 Monate alte Tiere).

Hinsichtlich der 15 Tage alten Versuchstiere der vorliegenden Untersuchungen kann allgemein festgestellt werden, daß Dunkelaufzucht die Spinesdifferenzierung an CA 1-Neuronen verzögert, während Dauerbeschallung eine Differenzierungsbeschleunigung bewirkt. Entsprechend liegen die Werte der Dunkeltiere zwischen den Angaben für 10 und 15 Tage alte Normaltiere, die der beschallten Tiere zwischen den Angaben für 15 und 20 Tage alte Normaltiere (vgl. Tab. 4, Abb. 2 und 4).

Tabelle 4. Spinesanzahl, Spinesdichte und Dendritenlänge in den Einzeldendritenfeldern der Basaldendriten, der apikalen Seitendendriten und des apikalen Büschels von CA 1-Neuronen bei Kontrolltieren, Dunkeltieren und beschallten Tieren.
(arithmetische Mittelwerte und Standardabweichungen)
Abkürzungen s. Tab. 1

		Basal-dendriten	apikale Seiten-dendriten	apikales Büschel
Kon-troll-tiere	S	$82,2 \pm 70,1$	$30,1 \pm 26,9$	$46,9 \pm 29,9$
	SD	0,26	0,36	0,34
	DL	$321,3 \pm 229,7$	$90,7 \pm 78,2$	$163,5 \pm 129,7$
Dunkel-tiere	S	$47,3 \pm 35,1^{++}$	$17,1 \pm 15,4^{++}$	$26,4 \pm 21,8^{+}$
	SD	$0,17^{++}$	$0,2^{++}$	$0,18^{++}$
	DL	$286,8 \pm 206,8^{-}$	$90,7 \pm 69,2^{-}$	$149,8 \pm 141,5^{-}$
be-schallte Tiere	S	$114,1 \pm 94,6^{+}$	$36,4 \pm 31,4^{+}$	$57,5 \pm 53,4^{-}$
	SD	$0,33^{++}$	$0,4^{+}$	$0,43^{+}$
	DL	$350,4 \pm 323,1^{-}$	$94,3 \pm 72.5^{-}$	$148,8 \pm 122,4^{-}$

Besonders deutlich treten die Unterschiede zwischen den drei Tiergruppen der vorliegenden Untersuchung bei Betrachtung der mittleren Gesamtspinesanzahl eines CA 1-Neurons hervor.[1]) Während eine CA 1-Pyramidenzelle bei den Kontrolltieren im Mittel 818 Spines besitzt, beträgt der Wert für die Dunkeltiere nur 478 Spines (—41,5%). Deutlich über dem Kontrolltierwert liegen die Angaben für ein durchschnittliches Neuron bei den beschallten Tieren mit 1029 Spines (+25,8%).

Diskussion

Mehrfach reproduzierte Untersuchungen an einem umfangreichen Tiermaterial haben gezeigt, daß eine Umweltreizanreicherung (enriched conditions and training — ECT) zu einer Zunahme des Hirngewichtes, der Cortexdicke und der Zellzahl in verschiedenen Regionen des Gehirns führt (Rosenzweig et al. 1962, 1969; Diamond et al. 1964, 1966, 1967; Walsh et al. 1969; Uylings 1977). Es stellt sich die Frage, welche Prozesse den genannten quantitativen Veränderungen, die offensichtlich durch einen größeren sensorischen Input induziert werden, zugrunde liegen. Da die Dickenzunahme der Rindenschichten in den untersuchten Regionen mit einem erhöhten Verzweigungsgrad dendritischer Fortsätze einhergeht (Holloway 1966; Uylings 1977) und da der Anstieg in der Zellzahl insbesondere auf eine Zunahme an Gliazellen zurückzuführen ist (Diamond et al. 1966; Walsh et al. 1969), kann vermutet werden, daß

1. erhöhter sensorischer Input mit einer Zunahme rezeptiver Strukturen verbunden ist, wobei
2. die damit entstehenden größeren nutritiven Anforderungen durch die Zunahme gliöser Elemente gedeckt werden könnten (Walsh et al. 1969).

1 Die für diese Berechnungen erforderlichen Daten für den apikalen Hauptdendriten wurden aus Frotscher et al. (1975) übernommen.

Durch zahlreiche Untersuchungen an dendritischen Spines wurde bestätigt, daß Umweltreizanreicherung zu einer Vermehrung und Umweltreizverarmung zu einer Abnahme bzw. Unterentwicklung rezeptiver Strukturen führt (VALVERDE 1967, 1971; SCHAPIRO and VUKOVICH 1970; FIFKOVÁ 1970; PARNAVELAS et al. 1975; FROTSCHER et al. 1975 u. a.). Da die dendritischen Spines das entscheidende postsynaptische Strukturelement für corticale Neurone darstellen (SCHEIBEL and SCHEIBEL 1970), induziert vermehrter sensorischer Input offensichtlich einen höheren Grad der synaptischen Verschaltung. Die vorliegenden Untersuchungen liefern weiteres Material für diese Auffassung. Umweltreizanreicherung (Dauerbeschallung) resultiert in einer Zunahme dendritischer Spines an apikalen Seitendendriten, Basaldendriten sowie an Dendriten des apikalen Büschels von Pyramidenneuronen der CA 1-Region des Hippocampus. Gegensinnige Effekte wurden durch Umweltreizverarmung (Dunkelaufzucht) erzielt. Unter Hinzunahme der Ergebnisse am apikalen Hauptdendriten (FROTSCHER et al. 1975) wird gefolgert, daß die entsprechenden Umweltfaktoren die Differenzierung des Gesamtneurons beeinflussen; Hinweise für eine Bevorzugung bestimmter Dendritenabschnitte ließen sich nicht finden, da die Veränderungen in der Spineszahl gleichsinnig über die einzelnen Ordnungen an apikalen Seitendendriten und Basaldendriten auftreten und in den verschiedenen Segmenten des apikalen Hauptdendriten nachweisbar sind. Die vorliegenden Ergebnisse werden — insbesondere hinsichtlich der apikalen Seitendendriten — durch elektronenmikroskopische Untersuchungen zur Synapsenanzahl im Stratum radiatum der CA 1-Region bei unter identischen Bedingungen aufgezogenen Versuchstieren bestätigt. Dauerbeschallung bis zum 15. Postnataltag führt zu einer Zunahme der Synapsen im Neuropil des Stratum radiatum; Dunkelaufzucht dagegen resultiert in einer verminderten Synapsenanzahl (FROTSCHER et al. 1977). Allerdings ist es — besonders angesichts der z. T. sehr unterschiedlichen Ergebnisse von Spineszählungen in verschiedenen ZNS-Regionen nach Dunkelaufzucht — nicht angebracht, verallgemeinernd zu schlußfolgern, daß sensorische Deprivation in allen ZNS-Regionen, bei den verschiedenen Zelltypen und in verschiedenen Altersstufen immer einen Differenzierungsrückstand bewirkt (vgl. RYUGO et al. 1975; MEYER et al. 1976; FRIMMEL 1977; SCHÖNHEIT und SCHULZ 1977b u. a.). Die funktionelle Überbeanspruchung anderer Analysatoren bei Dunkelaufzucht kann sogar zu einer Spinesvermehrung in entsprechenden Hirnregionen führen (z. B. Spinesvermehrung in der Hörrinde nach Dunkelaufzucht: RYUGO et al. 1975).

Der Hippocampusformation wird eine besondere Bedeutung bei der Verarbeitung sensorischer Afferenzen in komplexe corticale Prozesse zugesprochen (ADEY 1967; SMYTHIES 1967; WENZEL et al. 1973). WALSH et al. (1969) fanden eine signifikante Erhöhung der Gliazellzahl und der Dicke der Hippocampusformation bei Tieren, die einer reizstarken Umwelt ausgesetzt waren. Unter Hinzunahme der Befunde der vorliegenden Untersuchungen kann somit gefolgert werden, daß Veränderungen der äußeren Umwelt zu Veränderungen der zellulären und synaptischen Organisation der Hippocampusformation führen, die das Korrelat veränderter Funktion, d. h. Informationsverarbeitung, sein dürften.

In den vorliegenden Untersuchungen wurden keine signifikanten Unterschiede zwischen den Versuchstiergruppen in den Dendritenlängen — sowohl in den einzelnen Ordnungen als auch im Einzeldendritenfeld — festgestellt. Auch die Häufigkeitsverteilung der einzelnen Dendritenordnungen, die ein Kriterium für das Verzweigungsmuster darstellt, war unverändert (Abb. 3). GREENOUGH and VOLKMAR (1973) und UYLINGS (1977) fanden dagegen Wachstumsveränderungen an Basaldendriten neocorticaler Pyramidenzellen nach Einwirkung von Umweltreizen. In den vorliegenden Untersuchungen deutet sich lediglich bei den Basaldendriten der Dunkeltiere ein Wachstumsrückstand gegenüber den beiden anderen Tiergruppen an (Abb. 2, Tab. 2 und 4), statistisch signifikante Unterschiede ließen sich jedoch nicht nachweisen. Demnach erscheinen synaptische Strukturen als das plastische Element im Differenzierungsprozeß der CA 1-Neurone. Allerdings läuft die Differenzierung des Dendritenbaumes zeitlich vor der Spinesdifferenzierung ab (MINKWITZ 1976), so daß in der Postnatalphase einsetzende Umweltfaktoren weniger Einfluß haben dürften. Von Untersuchungen an älteren Tieren ist bekannt, daß auch die Synapsen- bzw. Spinesdifferenzierung in späteren Ontogenesestadien weniger beeinflußbar ist (VALVERDE 1971; SMIRNOV 1972; VRENSEN and DE GROOT 1974; SCHÖNHEIT und SCHULZ 1977a; FRIMMEL 1977; FROTSCHER et al. 1977).

Da veränderte Umweltbedingungen veränderten sensorischen Input bedingen, stützen die vorliegenden Untersuchungen Hypothesen über eine afferenzabhängige Spinesdifferenzierung (HÁMORI 1973a, b). Weiterführende Erkenntnisse über die funktionsabhängige Differenzierung synaptischer Strukturen lassen sich unseres Erachtens durch direkte Stimulierung oder Ausschaltung afferenter Systeme erlangen. So konnte nachgewiesen werden, daß Läsionen der entorhinalen Rinde über einen transneuronalen Effekt nicht nur eine Minderdifferenzierung in der Anzahl, sondern auch in der Größe von Spines an CA 3-Neuronen bewirken (FROTSCHER, HÁMORI and WENZEL 1977). Elektrische Stimulation der entorhinalen Rinde dagegen resultiert in einer Größenzunahme der Spines der Molekularschicht des Gyrus dentatus (FIFKOVÁ and VAN HARREVELD 1977).

Literatur

ADEY, W. R.: Hippocampal states and functional relations with cortico subcortical systems in attention and learning. Brain Res. **27**, 228–245 (1967).

ARŠAVSKIJ, I. I., M. B. BERKINBLUT, N. D. VEDENSKAJA, L. S. ENEDENKO, S. A. KOVALEV, V. V. SMOLJANIKOV und L. M. CAJLACHJAN: Funktionelle Bedeutung der strukturellen Besonderheiten der Dendriten (russ.). Symposium Dendriten. Akad. Med. Nauk SSSR Moskva 1971.

BOGOSLOVSKAJA, L. S., I. A. LJUBINSKIJ, N. V. POZIN, E. V. PUCILO, L. A. SMELEV und T. M. SURA-BURA: Die Rolle der Varikositäten und Verzweigungen der Dendriten in der Informationsverarbeitung durch das Neuron (russ.). Symposium Dendriten. Akad. Med. Nauk SSSR Moskva 1971.

CAJAL, S. R. y: Studien über die Hirnrinde des Menschen. Aus d. Span. übers. von J. BRESLER. Heft 5: Vergleichende Strukturbeschreibung und Histogenesis der Hirnrinde. Anat. physiol. Betrachtungen über das Gehirn. Struktur der Nervenzellen des Gehirns. Leipzig 1906.

COLEMAN, P. D. and A. H. RIESEN: Environmental effects on cortical dendritic fields I. Rearing in the dark. J. Anat. (Lond.) **102**, 363–374 (1968).

DIAMOND, M. C., D. KRECH and M. R. ROSENZWEIG: The effects of an enriched environment on the histology of the rat cerebral cortex. J. Comp. Neurol. **123**, 117–119 (1964).

DIAMOND, M. C., F. LAW, H. RHODES, B. LINDNER, M. R. ROSENZWEIG, D. KRECH and E. L. BENNETT: Increases in cortical depth and glia numbers in rats subjected to enriched environment. J. Comp. Neurol. **128**, 117–126 (1966).

DIAMOND, M. C., B. LINDNER and A. RAYMOND: Extensive cortical depht measurements and neuron size increases in the cortex of environmentally enriched rats. J. Comp. Neurol. **131**, 357–364 (1967).

EAYRS, J. T.: The cerebral cortex of normal and hypothyroid rats. Acta Anat. **25**, 160–183 (1955).

ENGLISCH, H.-J., G. KUNZ und J. WENZEL: Zur Spines-Verteilung an Pyramidenneuronen der CA 1-Region des Hippocampus der Ratte. Z. mikrosk. anat. Forsch. **88**, 85–102 (1974).

FIFKOVÁ, E.: The effect of unilateral deprivation on visual centers in rats. J. Comp. Neurol. **140**, 431–438 (1970).

FIFKOVÁ, E. and A. VAN HARREVELD: Long-lasting morphological changes in dendritic spines of dentate granular cells following stimulation of the entorhinal area. J. Neurocytology **6**, 211–230 (1977).

FRIMMEL, G.: Zur Entwicklung der Neuronenstruktur der Fascia dentata: Quantitativ-neurohistologische, ultrastrukturelle und experimentelle Untersuchungen. Promotion A an der Humboldt-Universität Berlin, 1977.

FROTSCHER, M.: Die postnatale Entwicklung corticaler Neurone und ihre Beeinflussung durch ein Trauma bei Rattus norvegicus B. J. Hirnforsch. **16**, 203–221 (1975).

FROTSCHER, M., B. MANSFELD und J. WENZEL: Umweltabhängige Differenzierung der Dendritenspines an Pyramidenneuronen des Hippocampus (CA 1) der Ratte. J. Hirnforsch. **16**, 443–450 (1975).

FROTSCHER, M., B. LÖSSNER und J. WENZEL: Quantitativelektronenmikroskopische Untersuchungen zur umweltabhängigen Differenzierung von Synapsen im Hippocampus (CA 1) der Ratte. Z. mikrosk.-anat. Forsch. (1977), im Druck.

FROTSCHER, M., J. HÁMORI and J. WENZEL: Transneuronal effects of entorhinal lesions in the early postnatal period on synaptogenesis in the hippocampus of the rat. Exp. Brain. Res. **30**, 549–560 (1977).

GREEN, J. D.: The hippocampus. Physiol. Rev. **44**, 561–608 (1964).

GREENOUGH, W. T. and F. R. VOLKMAR: Pattern of dendritic branching in occipital cortex of rats reared in complex environment. Exp. Neurol. **40**, 491–504 (1973).

HÁMORI, J.: Developmental morphology of dendritic postsynaptic specializations. In: Recent developments of neurobiology in Hungary. Ed. by K. LISSÁK Vol. **4**, 9–32 (1973a). Budapest: Akadémiai Kiadó.

HÁMORI, J.: The inductive role of presynaptic axons in the development of postsynaptic spines. Brain. Res. **62**, 337–344 (1973b).

HOLLOWAY, R. L.: Dendritic branching: some preliminary results of training and complexity in rat visual cortex. Brain Res. **2**, 393–396 (1966).

KUNZ, G., L. HOLZ, H.-J. ENGLISCH und J. WENZEL: Untersuchungen zur Spinesdichte und Spinesverteilung an apikalen und basalen Dendriten von Lamina-V-Pyramidenzellen des sensorischen Cortex der Ratte. J. Hirnforsch **5/6**, 389–399 (1974).

MEYER, G., R. FERRES und J. J. FEO: Quantitative Analyse des Einflusses visueller Stimulierung und Deprivation auf die Entwicklung der Dendritendornen verschiedener telenzephalischer Zentren. 71. Versammlung der Anatomischen Gesellschaft. Rostock/Warnemünde 1976, Abstracts p. 180.

MEISSNER, W. W.: Hippocampal functions in learning. J. psychiat. Res. **4**, 235–304 (1966).

MICHALSKI, A., R. PATZWALDT, E. SCHULZ und B. SCHÖNHEIT: Quantitative Untersuchungen an primitiven Pyramidenzellen in der vorderen cingulären Rinde der Ratte. J. Hirnforsch. **17**, 143–153 (1976).

MINKWITZ, H.-G.: Zur Entwicklung der Neuronenstruktur des Hippocampus während der prä- und postnatalen Ontogenese der Albinoratte. III. Mitteilung: Morphometrische Erfassung der ontogenetischen Veränderungen in Dendritenstruktur und Spinesbesatz an Pyramidenneuronen (CA 1) des Hippocampus. J. Hirnforsch. **17**, 255–275 (1976).

MOREST, D. K.: The growth of dendrites in the mammalian brain. Z. Anat. Entwickl. Gesch. **128**, 290–317 (1969).

PARNAVELAS, J. G. and H. B. M. UYLINGS: Continous illumination from birth affects the morphology of neurons in the visual cortex of rats. Exp. Brain. Res. **23**, 154 (1975).

ROSENZWEIG, M. R., D. KRECH, E. L. BENNETT and M. C. DIAMOND: Effects of environmental complexity and training on brain chemistry and anatomy: a replication and extension. J. comp. physiol. psychol. **55**, 429–437 (1962).

ROSENZWEIG, M. R., E. L. BENNETT, M. C. DIAMOND, SU-YU-WU, R. W. SLAGLE and E. SAFFRAN: Influences of environmental complexity and visual stimulation on development of occipital cortex in rat. Brain Res. **14**, 427–445 (1969).

Ryugo, D. K., R. Ryugo, A. Globus and H. Killackey: Increased spine density in auditory cortex following visual or somatic deafferentiation. Brain. Res. **90**, 143–146 (1975).

Schapiro, S. and K. R. Vukovich: Early experience effects upon cortical dendrites: A proposed model for development. Science **167**, 292–294 (1970).

Scheibel, M. E. and A. B. Scheibel: Of pattern and places in dendrites. Intern. Rev. Neurobiol. **13**, 1–26 (1970).

Schönheit, B. und E. Schulz: Der Einfluß sensorischer Stimulation auf die Lamina V-Pyramidenneurone des Gyrus cinguli der Ratte. Z. mikrosk. anat. Forsch. (1977a), im Druck.

Schönheit, B. und E. Schulz: Der Einfluß sensorischer Deprivation auf die Lamina V-Pyramidenneurone des Gyrus cinguli der Ratte. Z. mikrosk. anat. Forsch. (1977b), im Druck.

Smirnov, G. D.: Functional and structural aspects of plasticity. Ergeb. exp. Med. **10**, 11–18 (1972).

Smythies, J. R.: Brain mechanism and behaviour. Brain **90**, 697–706 (1967).

Stensaas, L. J.: The development of hippocampal and dorsolateral pallial regions of the cerebral hemisphere in fetal rabbits. IV. Fortyone millimeter stage, Intermediate lamina. J. comp. Neurol. **131**, 409–422 (1967).

Uylings, H. B. M.: A study on morphometry and functional morphology of branching structures, with applications to dendrites in visual cortex of adult rats under different environmental conditions. Dissertation an der Universität Amsterdam 1977.

Valverde, F.: Apical dendritic spines of the visual cortex and light deprivation in the mouse. Exp. Brain Res. **3**, 337–352 (1967).

Valverde, F.: Rate and extent of recovery from dark rearing in the visual cortex of the mouse. Brain Res. **33**, 1–11 (1971).

Vrensen, G. and D. De Groot: The effect of dark rearing and its recovery on synaptic terminals in the visual cortex of rabbits. A quantitative electron microscopic study. Brain Res. **78**, 263–278 (1974)

Vrensen, G. and D. De Groot: The effect of monocular deprivation on synaptic terminals in the visual cortex of rabbits. A quantitative electron microscopic study. Brain Res. **93**, 15–24 (1975).

Walsh, R. N., O. E. Budtz-Olsen, J. E. Penny, R. A. Cummins: The effects of environmental complexity on the histology of the rat hippocampus. J. Comp. Neurol. **137**, 361–365 (1969).

Wenzel, J., W. Kirsche, G. Kunz, H. Neumann, M. Wenzel und E. Winkelmann: Licht- und elektronenmikroskopische Untersuchungen über die Dendritenspines an Pyramidenneuronen des Hippocampus (CA 1) bei der Ratte. J. Hirnforsch. **13**, 387–408 (1973).

Wenzel, J., B. Mannsfeld, H.-J. Englisch und G. Kunz: Über die Dendritenspines von Pyramidenneuronen (Lamina III) der Area striata der Ratte. Z. mikrosk.-anat. Forsch. **87**, 505–524 (1973).

Adresse des Autors:

Dr. med. Michael Frotscher
Anatomisches Institut des Bereiches Medizin (Charité)
der Humboldt-Universität zu Berlin
DDR - 104 Berlin, Philippstr. 12

J. Hirnforsch. 19 (1978) 457–467

Department of Anatomy, School of Medicine, Iwate Medical University Morioka 020, Japan

Variations of the dog cerebral sulci, compared in particular with those of the cat

By

Koki KAWAMURA and Jumpei NAITO

With 7 figures

(Received December 6, 1977)

Key words: Variations, cerebral sulci, dog, cat

Summary: Various types of 15 sulci, ill on the lateral and 4 on the medial surface, of the dog cerebral hemisphere, were examined. Each sulcus has several formes and a sulcal pattern which shows the highest frequency is taken as a "standard" type. A standard map of the dog brain is presented based upon the present observation. Sulcal patterns of the dog cerebrum are in general as consistent as those of the cat brain. However, some sulci, i.e., a small branch of the suprasylvian sulcus, the suprasplenial and postsplenial sulci, the ectolateral, the genual, the prorean, and the eetosylvian sulcus, show variety of formes.
Variations of the dog sulci are discussed and compared with those of the cat. The comparison shows that the sulcal patterns which appear only in the dog are on the whole most variable.

Introduction

Of carnivores, dogs and cats are the most used experimental animals. While cats are widely used in anatomical and physiological studies of the central nervous system, dogs are more popular animals in the field of psychology, especially as regards behavior related to conditioned reflexes. Since hybrid dogs are more common in neuroanatomical laboratories than dogs of pure species, it is more useful for students doing research on the dog's brain to obtain knowledge of the hybrid dog rather than of the pure species.

One of us (KAWAMURA, 1971) has previously described the variations in the sulcal pattern of the feline brain. In the present study, we have attempted to do the same in the dog. The aim of the paper is to give details concerning the sulcal pattern, thus making identification of the cortical topography easier for students operating on the dog's brain.

Material and methods

Forty-three adult dogs were used in this investigation. The brains were dissected free from the skull, usually after perfusion with 10% formalin, and thereafter transferred to a 10% formalin solution for further fixation. The arachnoid was removed and the lateral and medial surfaces of both hemispheres were examined with particular attention to the sulcal pattern. The brains were after this photographed (fig. 1) and reproduced in drawings. No consideration was paid to species, age, sex and body weight differences.

Results

Fifteen sulci, 11 on the lateral and 4 on the medial surface of the hemisphere, were examined. Various types of sulcal patterns were observed for each sulcus. These were divided into several groups (s, a, b, c, ... in figs. 2–4) indicating the percentage of the frequencies. In this paper, the sulcus with the highest percentage was taken as the "standard" (s) type, and the others considered as variation (v) types. In the figures, the features of the sulci are diagrammatically illustrated by thick lines. Approximate positions of the s-type sulcus are indicated by dotted lines, and those of the neighboring s-type sulci by broken lines.

Since domestic, hybrid dogs were used in this study with no consideration to species differences, the question may be raised if the results would have been different if pure species, such as the terrier, pointer or bulldog, had been studied. For a more detailed consideration of this question the reader is referred to other publications (KUBOTA, 1956; HASEGAWA, 1959).

1. *The sylvian sulcus* (fig. 2) is a short posterodorsally running fissure which ends in the sylvian gyrus (72%, s-type). Frequently its dorsal end bifurcates (16.4%, a-type) and its posterior branch often joins the ectosylvian sulcus (11.6%, b-type). Thus, it belongs to the most stable type of sulci and its ventral portion, mostly separated, joins the rhinal fissure covering the insula.

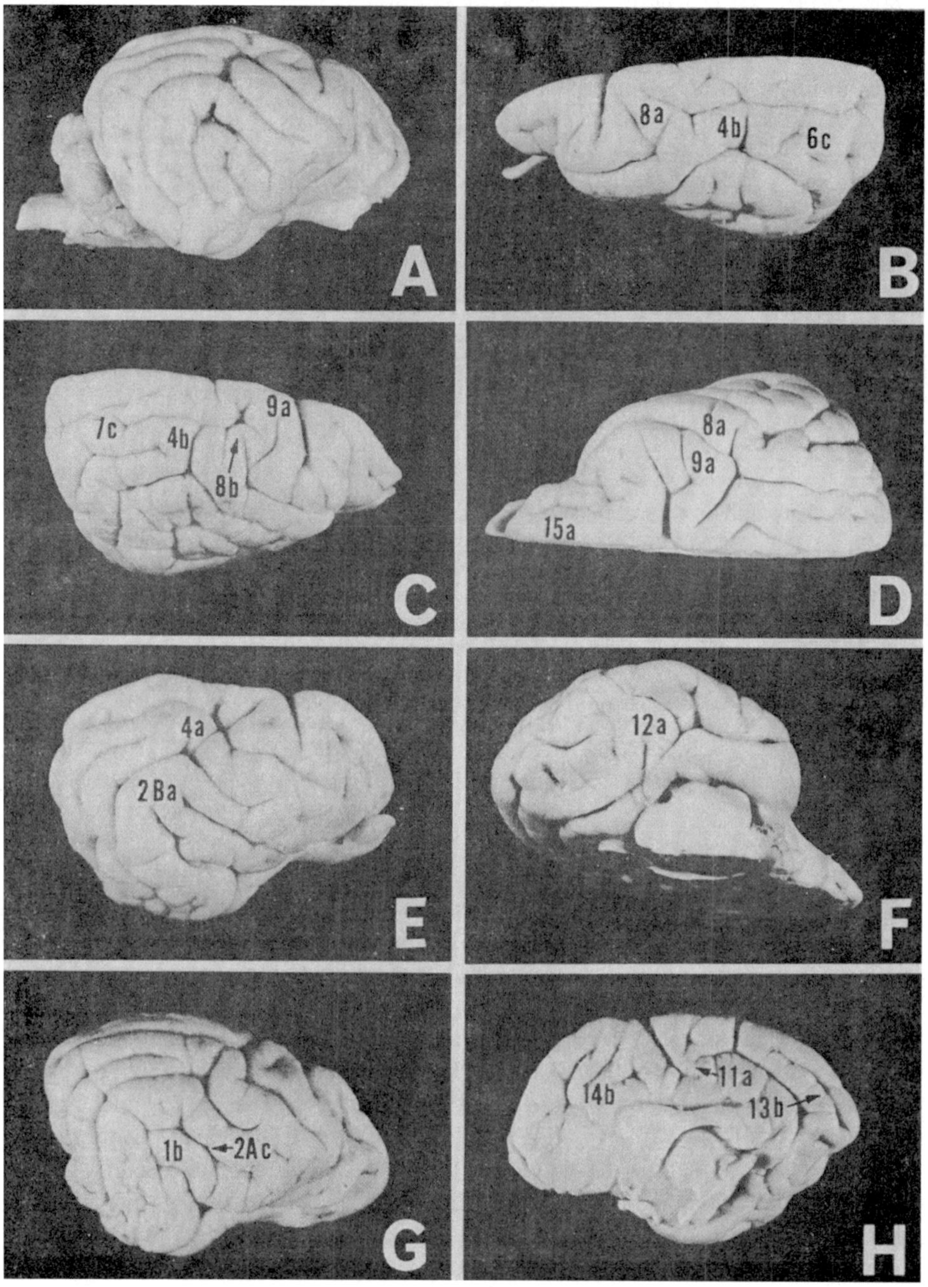

Fig. 1. Photos of 8 dog brains to illustrate various types of sulcal patterns. The indicated numbers are used in the text. To the left, 4 pictures of the lateral surface of the hemisphere are shown including a "standard" type (A). In these, a branch from the anterior flexure of the suprasylvian sulcus is seen to link with the lateral sulcus (4b in C) or with the ansate sulcus (4a in E). In photo G, formation of the sylvian gyrus is poor and the sylvian sulcus is linked with the ectosylvian sulcus (1b and 2Ac). The "cat gyrus" is observed in E (2Ba). At bottom right (photo H) a medial view is shown where the cruciate and splenial sulci are separated (11a), and divergent rami of the genual sulcus are conspicuous (14b). In the dorsal view (B and D), the cruciate, the postcruciate and the coronal sulci are seen to constitute a continuous line (9a in D), and the ectolateral sulcus is underdeveloped in B (6c).

Figs. 2–4. Schematic drawings of various sulcal patterns, indicating the percentage (in parenthesis) after each alphabetical letter. A letter "s" denotes a "standard" type whose frequency is the highest among the cases, the other letters represent types of the variants. The numbers to the left are the same as those in the text and figs. 1 and 5. The course of the sulcus under study is indicated by thick lines and approximate positions of the s-type sulcus are indicated by dotted lines, those of other sulci of s-types are shown by broken lines.

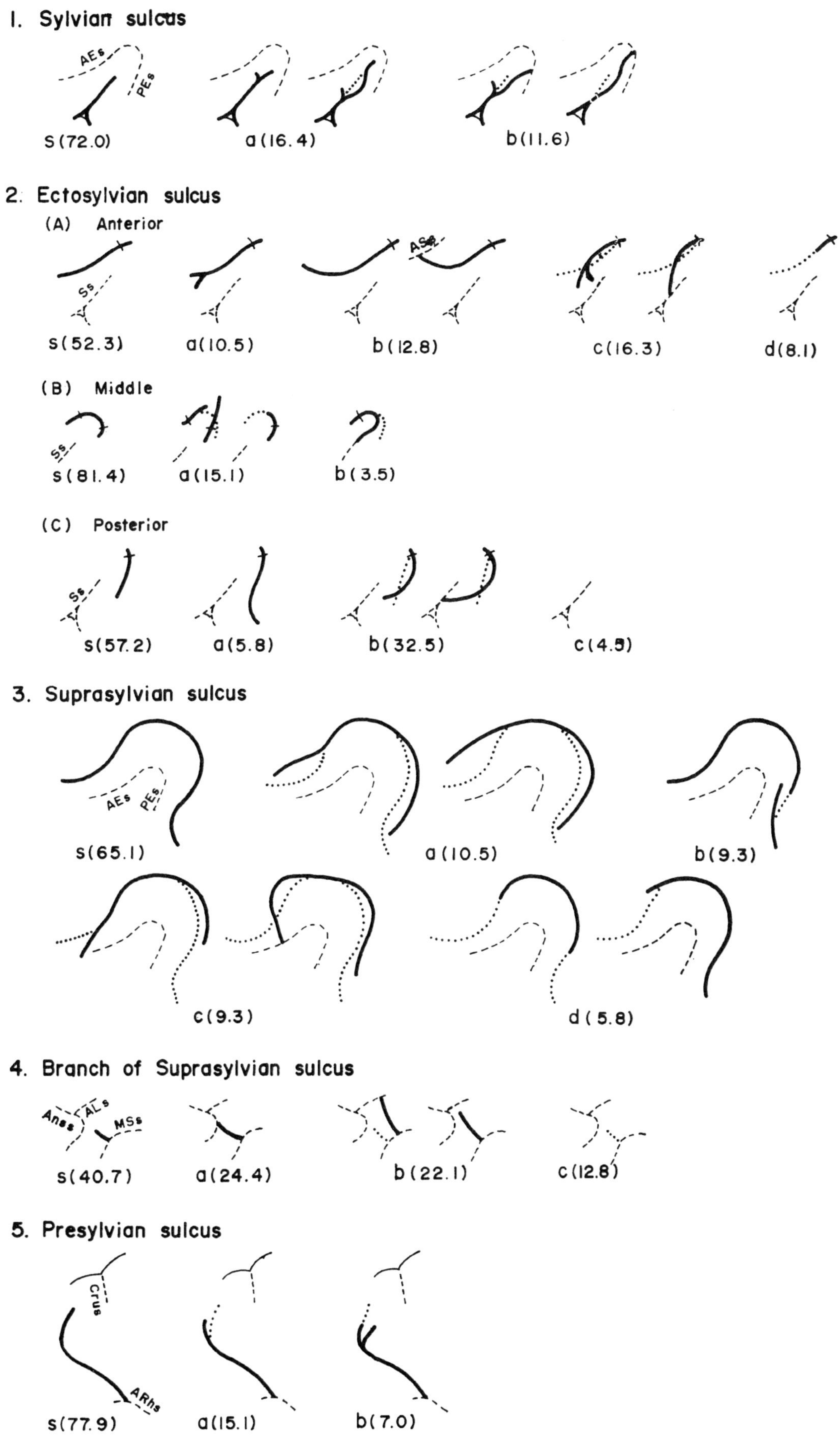
1. Sylvian sulcus
AEs
PEs
s(72.0)
a(16.4)
b(11.6)
2. Ectosylvian sulcus
(A) Anterior
Ss
ASs
s(52.3)
a(10.5)
b(12.8)
c(16.3)
d(8.1)
(B) Middle
Ss
s(81.4)
a(15.1)
b(3.5)
(C) Posterior
Ss
s(57.2)
a(5.8)
b(32.5)
c(4.5)
3. Suprasylvian sulcus
AEs
PEs
s(65.1)
a(10.5)
b(9.3)
c(9.3)
d(5.8)
4. Branch of Suprasylvian sulcus
Anss
ALs
MSs
s(40.7)
a(24.4)
b(22.1)
c(12.8)
5. Presylvian sulcus
Crus
ARhs
s(77.9)
a(15.1)
b(7.0)

2. *The ectosylvian sulcus* (fig. 2) separates the sylvian gyrus from the ectosylvian and it shows fairly variable patterns. Since the entire sulcus is divisible into three parts: (A) anterior, (B) middle and (C) posterior, a description of these is given separately below.

A. *The anterior ectosylvian sulcus* is deep and ends mostly (more than 60%) in the anterior portion of the temporal area. Its anterior part may run forward to reach or almost join either the anterior suprasylvian (12.8%, b-type), or the sylvian sulcus (16.3%, c-type). In rare cases (8.1%, d-type), it appears to be a remnant.

B. *The middle ectosylvian sulcus* is a stable (81.4%, s-type) portion of the entire length of the ectosylvian sulcus. However, some variable patterns can be found. Thus, it can be interrupted in the middle portion (15.1%, a-type), and its caudal end may join the sylvian sulcus (3.5%, b-type).

C. *The posterior ectosylvian sulcus* is not always complete at its lower end (see c-type, 4.5%), in such cases posterior sylvian and the ectosylvian gyri appear to be continuous. On the whole, however, it is well developed and sometimes it is inclined to reach the sylvian sulcus (32.5%, b-type) or the posterior suprasylvian sulcus (5.8%, a-type).

3. *The suprasylvian sulcus* (fig. 2) is a long arched cleft that surrounds the ectosylvian formation. Usually (84.9%, s-, a-, and b-types) it takes a forward course, but often curves down to join the ectosylvian sulcus (9.3%, c-type). The anterior suprasylvian sulcus is lacking in 5.8% (d-type), and discontinuity of the posterior suprasylvian sulcus is observed in 9.3% (b-type). In the a-type (10.5%), the sulcus is vaulting in the entire length.

4. *A forward branch of the suprasylvian sulcus* (fig. 2) springs from the anterior portion of the suprasylvian sulcus (87.2%), joining either the ansate (24.4%, a-type) or the lateral sulcus (22.1%, b-type).

5. *The presylvian sulcus* (fig. 2) is a long and deep furrow which runs as far as to the anterior sigmoid

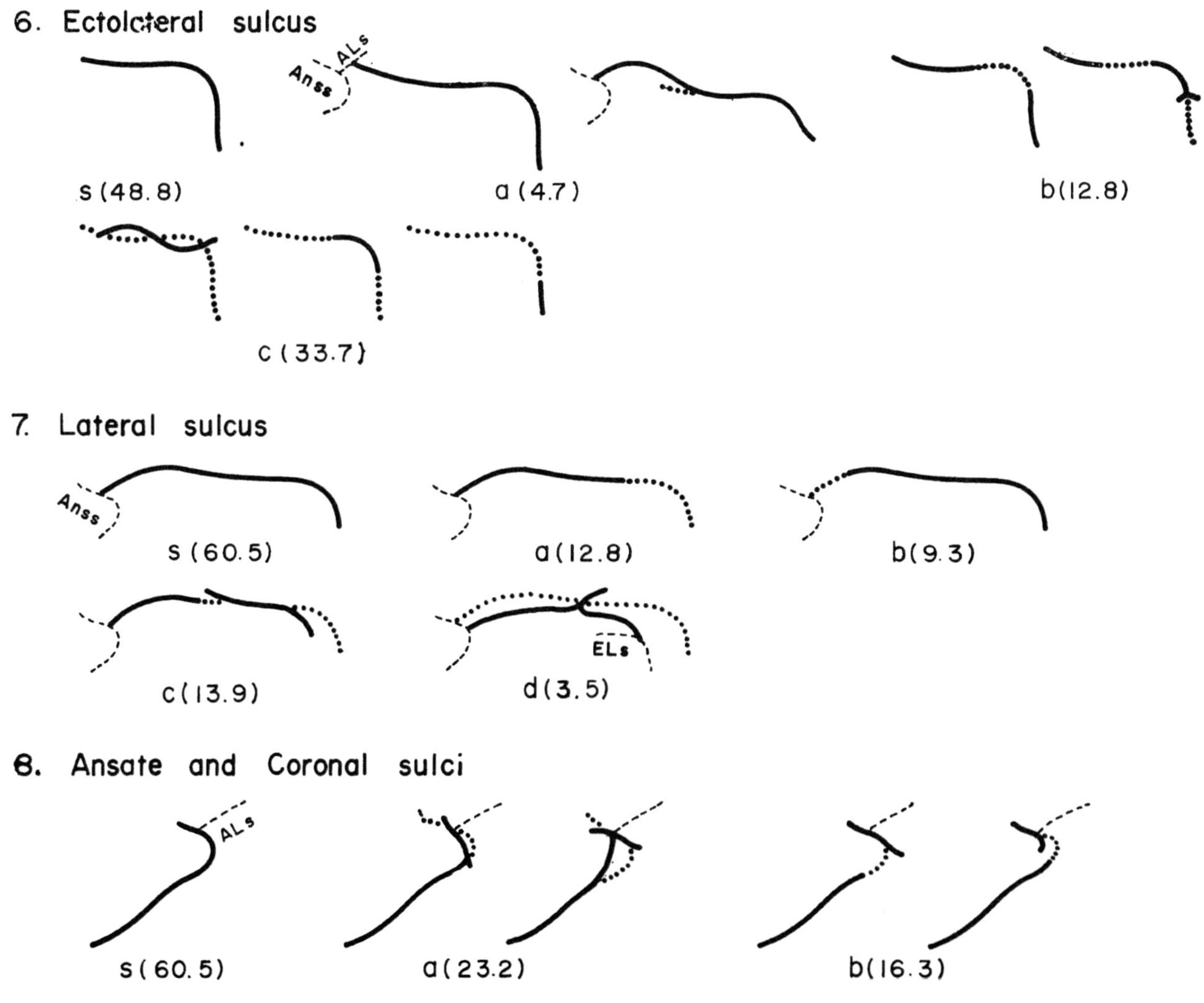

Fig. 3. For legend see Fig. 2.

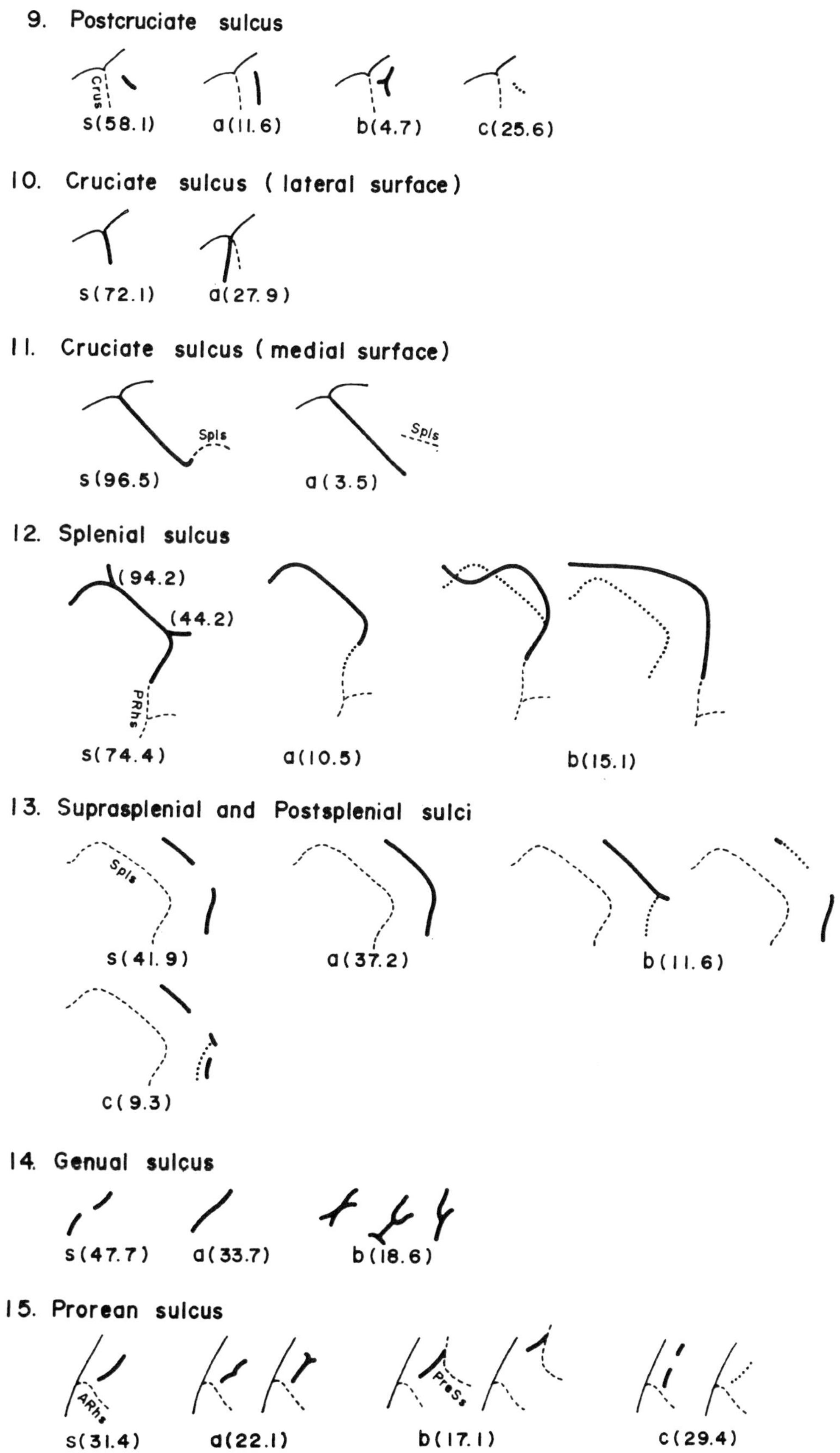

Fig. 4. For legend see Fig. 2.

gyrus (77.9%, s-type). Its course is stable, but sometimes it is shortened (15.1%, a-type) or bifurcated (7.0%, b-type).

6. *The ectolateral sulcus* (fig. 3) is a newly developed furrow that separates the suprasylvian gyrus from the ectolateral gyrus. It consists of the horizontal (or anterior) and the vertical (or posterior) branches, and it is highly unstable (note that only 48.8% is the "standard" type). Thus, it is poorly developed (33.7%, c-type) and may be disrupted into two, or even three parts (12.8%, b-type). Rarely (4.7%, a-type) it connects either with the lateral or with the ansate sulcus.

7. *The lateral sulcus* (fig. 3) runs in a parasagittal plane on the dorsal surface of the hemisphere as far back as the occipital pole. In a small percentage (9.3%, b-type) the lateral sulcus does not connect with the ansate, and sometimes (13.9%, c-type) it is discontinuous, or it lacks its posterior portion (12.8%, a-type). The posterior end of the lateral sulcus may join the ectolateral sulcus (3.5% in d-type). In some brains, another shallow furrow, the entolateral sulcus, is observed in the lateral gyrus.

8. *The ansate and the coronal sulci* (fig. 3). In the majority of cases (60.5%, s-type), the lateral branch of the ansate sulcus curves forward to connect the coronal sulcus, hence given the name of the corono-ansate sulcus (see Yoshikawa, 1962). Sometimes (23.2%, a-type) the coronal sulcus originates from the middle portion of the ansate. Thus the confluence of the two sulci is common and even if the separation of the sulci is observed in 16.3% (b-type), it appears that they are linked by a shollaw dimple.

9. *The postcruciate sulcus* (fig. 4) is usually a small and shallow furrow to be seen in the posterior sigmoid gyrus (58.1%, s-type). Sometimes it is more developed (11.6%, a-type), and bifurcated (4.7%, b-type). In 25.6% of the cases examined, it is lacking.

10. *The cruciate sulcus on the lateral surface* (fig. 4) is a fissure running in the sigmoid gyrus, thus dividing the gyrus into an anterior and a posterior portion. In most cases (72.1%, s-type), the lateral limb courses lateralward, while it may curve a little forward (27.9%, a-type).

11. *The cruciate sulcus on the medial surface* (*the medial limb*) (fig. 4) is continuous with the lateral limb on the lateral surface. In almost all cases examined (96.5%, s-type), the sulcus joins the splenial sulcus.

12. *The splenial sulcus* (fig. 4) is a deep cleft which separates the mesocortex from the neocortex. In the majority of cases, its anterior and posterior ends connect with the cruciate and the posterior rhinal sulcus, respectively. The connection with the cruciate sulcus is mentioned above. As for the confluence with the posterior rhinal sulcus, it is seen in 89.5% (s- and b-types). Two small branches from the splenial sulcus are frequently observed; one arises forward (94.2%) and the other backward (44.2%).

13. *The suprasplenial and the postsplenial sulci* (fig. 4) are both shallow longitudinal furrows running ventrally to the suprasplenial gyrus. In 41.9% (s-type), the two sulci are separated from one another. In some cases (11.6%, b-type), one of the two is missing.

14. *The genual sulcus* (fig. 4) is an irregularly curved, shallow dimple located in the medial surface of the hemisphere in front of the corpus callosum. Patterns of its configuration vary. In the majority of cases it has an interruption in its middle part (47.7%, s-type), while in 33.7% (a-type) it shows a rather long curve which separates the frontal gyrus from the ventrally situated gyrus rectus. A third type with divergent (18.6%, b-type) rami can be observed.

15. *The prorean sulcus* (fig. 4), situated on the lateral surface of the frontal cortex, has various types of configuration. In 29.4% (c-type) it is poorly developed, or lacking. This variability may in part be associated with the course of the presylvian sulcus, with which the prorean sulcus may be connected (17.1%, b-type).

Discussion

Various types of sulcal patterns of the domestic, hybrid dog are described in this study. They are diagrammatically presented in the Results. A standard, composed map of the dog brain based upon the detailed observations of the present study is illustrated in figure 5.

A comprehensive evaluation of the cause of the variability of the sulcal pattern is beyond the scope of the present study. However, before entering into a discussion of the pattern of the canine sulci and a comparison of these with the feline type, developmental and functional aspects of the dog cerebrum will briefly be considered.

1. Developmental aspects of the canine brain

Postnatal development of the dog cerebrum does not proceed equally for all parts; on the contrary, certain local predilections are obvious (Herre and Stephan, 1955). Thus, the development of the rhinal as well as

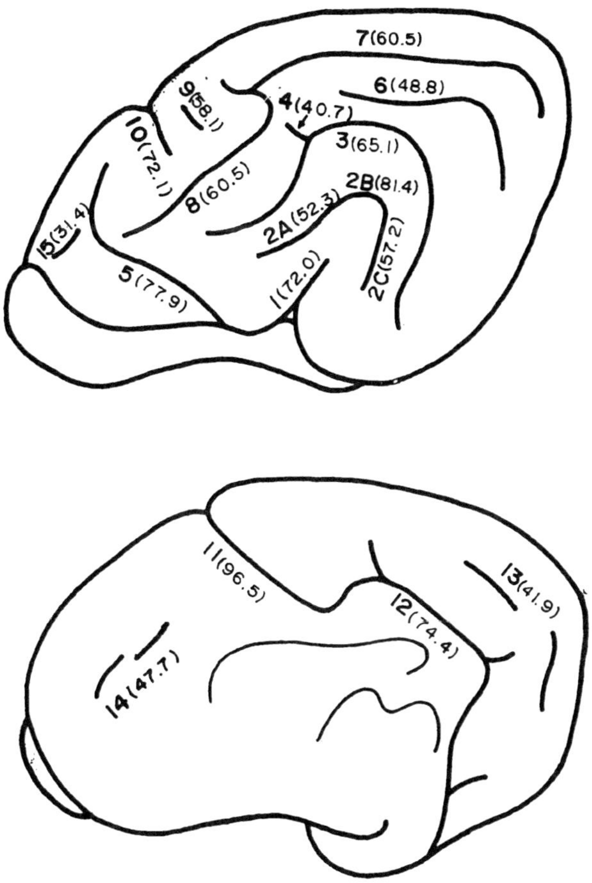

Fig. 5. A composite map of the dog brain with the standard type of sulci indicated with numbers (from 1 to 15) and frequencies (percentages in parentheses).

the basal part of the frontal areas progresses rostrodorsally; the caudal part of the temporal area turns rostrally, and the occipital region grows from the medial surface to the lateral. In addition, the basal part of the brain is pushed forward by growth of the cerebellum. These developmental processes bring about some significant effects on the sulcal formation of the entire brain. First, the sulci of the sylvian system are forced ventrorostrally, and second, the anterior and posterior limbs of the ectosylvian as well as the suprasylvian sulci are strongly pressed toward the sylvian sulcus. Third, the coronolateral, the ectolateral and the suprasylvian sulci, which at birth occur as gentle arches, become more dorsally convex. It is noteworthy that RADINSKY (1973), in his phylogenetical study of endocranial casts of fossil and recent carnivora, has demonstrated a similar type of evolution of the canine brain. For example, according to RADINSKY (1973), an interruption between the anterior and posterior limbs of the *ectosylvian sulcus* existed in the ancient dog of 25 million years ago, the limbs becoming united much later (12—14 million years ago). Therefore, the separation of the sulcus which is observed in our present materials in 15.1% (fig. 2) may indicate the phylogenetically old pattern. In other carnivores such as the Ungulata and the Mustelida, a typical, well formed ectosylvian sulcus can not be observed (BOLK et al., 1934). The *suprasylvian sulcus* is, on the other hand, phylogenetically old since this can be seen in the fossils of ancient dogs as well as in Marsupialia and primitive Ungulata (Hyrax) (BOLK et al., 1934; RADINSKY, 1973). The *ectolateral sulcus* to appear between the suprasylvian and the lateral sulci is phylogenetically new and shows highly variable patterns. Thus, the sulcus is rudimentary in the ancient dogs and develops into a long sulcus in the modern forms (RADINSKY, 1973). The *presylvian sulcus*, located in the frontal cortex, is one of the most stable sulci, its form being nearly accomplished at birth (HERRE and STEPHAN, 1955). In contrast to this, the patterns of the *genual* and *prorean sulci* are extremely variable.

2. *Cortical functional areas in the dog cerebrum*

In 1944 and 1945, the *auditory cortex* of dogs was indicated by TUNTURI in the ectosylvian gyrus and in a small dorsal portion of the sylvian gyrus. For his identification, he used audio-frequency signals and electrical stimulation of cochlear nerve fibers. Subsequently, with the strychnine stimulation method the same author (1950) argued that the dorsal part of the sylvian gyrus should not be included in the auditory area, and that failure to locate the area in the dog was due either to an anomaly of the acoustic system or to the fact that the region still remained hidden within the ectosylvian sulcus. From a study of retrograde cell degeneration, however, SYCHOWA (1963) stated that the auditory area of the dog locates in other areas than the ectosylvian gyrus, since cortical ablation of the anterior and posterior sylvian gyri produces explicit cellular degeneration in a definit region of the medial geniculate body. The discrepancy among the authors as to the location of the auditory cortical area may largely be due to the variability of the dorsal part of the sylvian gyrus, shown in the present investigation (fig. 2). In addition, TUNTURI (1945) located the third auditory area (A III) in a restricted ventral part of the dog anterior ectosylvian gyrus. Its location appears to be similar in the cat (HAMUY et al., 1956). However, this statement has been disputed (WOOLSEY, 1960).

The *visual cortical area* of the dog appears not to have been delineated physiologically. In the cytoarchitectonic map of GUREWITSCH and BYCHOWSKY (1928), the visual cortex is indicated, on the lateral surface of the hemisphere, in the posterior part of the anterior lateral gyrus as well as in the posterior

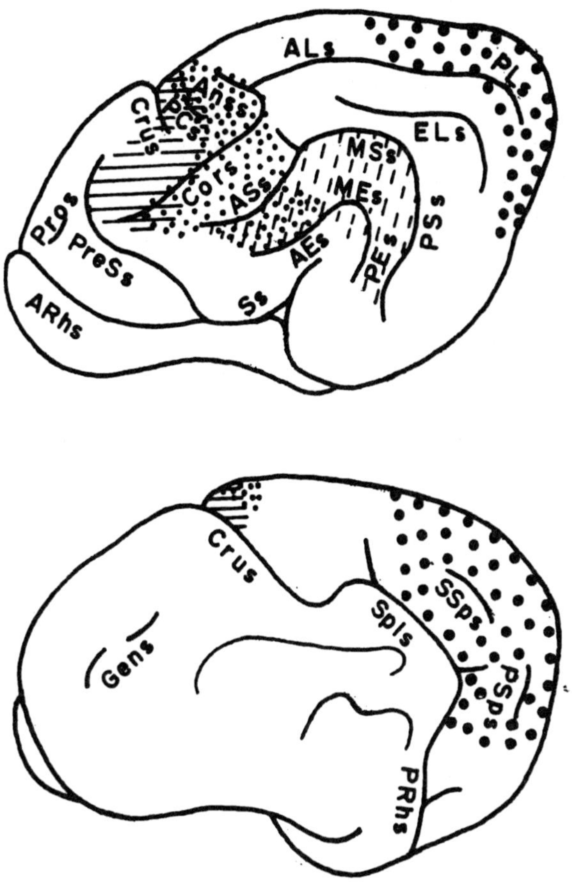

Fig. 6. A diagram (slightly modified) to show various functional areas of the dog cortex on a standard map based upon the findings of HAMUY et al. (1956) (the somatic sensory area, indicated by small dots), of GUREWITSCH and BYCHOWSKY (1928) (the visual area, large dots), of TUNTURI (1950) (the auditory area, vertical broken lines) and of GÓRSKA (1974) (the motor area, horizontal lines). Note areas of overlap between the somatic sensory and motor areas in the sigmoid and coronal gyri, and also those of overlap between the somatic sensory and the audotiry areas in the anterior ectosylvian gyrus. The lateral surface above, the medial below. Names of sulci are indicated by abbreviations.

lateral gyrus, and on the medial surface, in areas situated posterodorsally to the splenial sulcus. Since our observations indicate that the ectolateral sulcus is highly unstable and in more than a third of the cases are incomplete, the question whether or not this sulcus represents the boundary line of the visual area is obviously not solved.

With the evoked potential technique, the pattern of representation for the first and second *somatic sensory areas* (S I and S II) was mapped by HAMUY et al. (1956), and for the motor cortex (M I and M II) by GÓRSKA (1974). In the map of HAMUY et al. (1956), the S I is shown as a continuous band located in the coronal and posterior sigmoid gyri, while the S II is placed in the anterior ectosylvian gyrus. From their figure 2, the ansate sulcus and a small branch of the suprasylvian sulcus appear to represent the posterior border of S I. This largely corresponds with the cytoarchitectonical studies of CAMPBELL (1905) and GUREWITSCH and BYCHOWSKY (1928). The S II of HAMUY et al. (1956), on the other hand, largely overlaps the anterior portion of the ectosylvian B of CAMPBELL (1905) or the anterior part of area 50 in the map of GUREWITSCH and BYCHOWSKY (1928).

The functional organization of the dog *motor cortex* was studied by GÓRSKA (1974) by means of electrical stimulation. Based upon observations of somatic movement elicited with different levels of threshold, she differentiated the "precentral" motor area (M I) in the anterior half of the posterior sigmoid gyrus and the supplementary motor area (M II) in the lateral two-thirds of the anterior sigmoid gyrus, indicating the former area to overlap to some extent with the S I, stimulation of which could also elicit somatic movement with much higher threshold than in cases of stimulation of the motor proper area. The M I corresponds approximately to area 4a of KLEMPIN (1921), while area 4 of GUREWITSCH and BYCHOWSKY (1928) extends further to the anterior sigmoid and coronal gyri. The M II, on the other hand, extends from area 4aβ to the ventral part of area 6a of KLEMPIN's map (1921).

In figure 6, the cortical functional areas in the dog as discussed above are illustrated with different symbols.

3. *Course of cerebral arteries*

The arteries covering the cerebral surface are meandering on the gyral surfaces and enter the sulci, accompanying the pia. They ramify therein and come out again from different portions of the sulci. Thus, the course of the arteries does not coincide with that of the sulci. Figure 7 schematically illustrates the course of the dog cerebral arteries observed in our material. It is evident from the diagram that the main arteries do not always travel along deep sulci.

The relationship between the arterial blood flow and the sulcus formation in the gyrencephalic brain may be an important issue and may need a special developmental study. Phenomenologically viewed, however, it might be alleged that the arterial flow will not be an "active" factor in forming the cerebral sulci.

4. *Comparison with sulcal variations of the cat*

The cerebral convolution of the lateral hemisphere of the gyrencephalic brain can generally be divided into 1) the sylvian (the gyri lying ventral to the

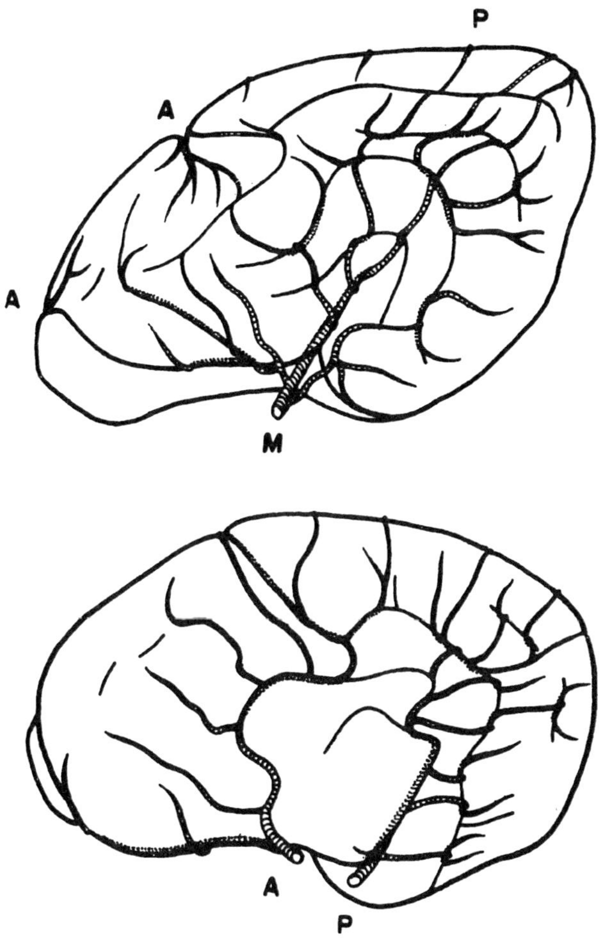

Fig. 7. Diagrams of the lateral (above) and the medial (below) views of the dog cerebral hemisphere to show the course of the anterior (A), the middle (M) and the posterior (P) cerebral arteries. Smaller branches are not indicated here.

suprasylvian sulcus) and 2) the marginal system (the gyri situated between the suprasylvian and splenial sulci), the latter being phylogenetically older than the former (Bolk et al., 1934). The various patterns of the cerebral sulci in the dog as revealed in the present investigation can be grouped into 4 systems: (A) the sylvian, (B) the marginal, (C) the splenial and (D) the others. Some morphological features will be discussed below with particular reference to those of the cat (cp. Kawamura, 1971).

A. The sylvian system (the sylvian, ectosylvian, suprasylvian and presylvian sulci)

The *ectosylvian* sulcus represents a characteristic difference between dogs and cats. Thus, while the sulcus is continuous in the dog, it is divided into two parts in the cat, forming the "cat gyrus" described by Meynert in 1877. We observed, however, a separation in 15.1% even in the dog (cp. also data of Filimonoff, 1928, noted 11%). In the cat the anterior and posterior ectosylvian sulci are inclined to join in 15.1% (Kawamura, 1971).

The course of the *suprasylvian* sulcus is more stable in the cat than in the dog. In the latter, a small branch from the anterior flexure of the sulcus occurs in 87.2%, while seldom in the cat (8.4%, a-type of Kawamura, 1971). The branch may represent a posterior boundary of the somatic sensory area (Campbell, 1905; Gurewitsch and Bychowsky, 1928; Hamuy et al., 1956).

Because of the well developed prorean gyrus, which largely corresponds to areas 6 and 8 of Klempin (1921) and of Gurewitsch and Bychowsky (1928), the *presylvian* sulcus in the dog takes a longer course than in the cat. This relatively larger development of the canine frontal cortex may reflect a morphological basis for the accessibility to behavior training of this animal.

B. The marginal system (the lateral, ectolateral, postcruciate, ansate and coronal sulci)

The *lateral sulcus* in the dog breaks up in its middle protion only in 13.9% (c-type) and the "bridge formation" between the anterior and the posterior branches of the lateral sulcus is much more frequently (42.8%, see Kawamura, 1971) observed in the cat. The *sulcus accessorius intralateralis,* (type II of Otsuka and Hassler, 1962) is found running medially to the cat lateral sulcus in 9.5%, while it is not obvious in the dog. Instead, the *ectolateral* sulcus, situated laterally to the lateral sulcus, is seen in most cases in the dog, although sometimes (over 30%) it is underdeveloped and even rudimentary. The present authors are of the opinon that the two sulci are of different origin, and that the one in the cat is located in the visual area while that in the dog is situated in the "association" cortex.

Although the *postcruciate* sulcus is missing in 59.0% in the cat hemisphere (Kawamura, 1971), it is well developed in the dog, and is seen in 74.4% of the total cases examined (cp. Filimonoff, 1928, noted 96%). The *ansate* and the *coronal* sulci are in most cases (83.7%) continuous in the dog and a single branch deviates frequently (19.8%) toward the postcruciate sulcus from approximately a posterior third portion of the coronal sulcus. We observed one case where the coronal sulcus was linked to the postcruciate and further to the cruciate sulcus (Photo D in fig. 1). On the basis of myeloarchitectonic as well as cytoarchitectonic studies, Kreiner (1961) presumed this sulcal line to be homologous to the central sulcus of the primate brain. Physiological

(Hamuy et al., 1956) as well as morphological (Campbell, 1905; Klempin, 1921; Gurewitsch and Bychowsky, 1928) data may support this view. On the other hand, it is generally considered that in the cat the somatic sensory and motor areas can not be clearly separated by certain sulci but overlap to some extent (see, e.g., Kusama et al., 1966; Nakahama, 1961). The same is obviously the case for the dog brain, since somatic movements can be elicited by stimulation of the somatic sensory area (Górska, 1974).

C. The splenial system (the cruciate, splenial, supra- and post-splenial sulci)

On the medial surface of the dog hemisphere, the *cruciate* and *splenial* fissures blend to form a long, continuous furrow (96.5%) to join the posterior rhinal sulcus (89.5%). However, in the cat, the splenial sulcus is neither continuous with the cruciate (79.2%, from Kawamura, 1971) nor with the postrhinal sulcus. In addition, the *suprasplenial* and the *post-splenial* sulci are continuous in 37.2% in the dog, but only in 14.6% (Kawamura, 1971) in the cat.

D. The prorean and the genual sulci

The two sulci are located in the frontal cortex and they are extremely variable in form. The *prorean* sulcus can be seen only in the dog (over 70%), presumably as the result of the advanced development of the frontal cortex in this animal. The *genual* sulcus lies in the vicinity of the prorean sulcus, but located on the medial surface of the brain. The sulcus in the dog is much more developed than in the cat. In the latter animal, it lacks in 56.2% of the cases (Kawamura, 1971).

From the results of the present study, the dog cerebral sulci can be divided into 3 categories: the stable, the variable and the most variable. The sylvian, the presylvian, the cruciate and the splenial sulci belong to the stable group (over 70% of s-type). In the most variable group (frequency of the s-type is below 50%), a small branch of the surpasylvian sulcus, the ectolateral, the surpa- and the post-splenial, the genual and the prorean sulci may be included. The rest of them, whose frequency of the s-type ranges from 50% to 70%, may belong to the variable group.

Acknowledgements

Part of the present material has been provided from Mr. K. Miyama and Dr. S. Yagi, to whom we are indebted. We thank Drs. T. Matsuno and G. Kodama for their discussions when a preliminary report (Naito and Kawamura, 1975) of this study was presented in a local anatomical meeting in Morioka.

List of abbreviations

A: Anterior cerebral artery
AEs: Anterior ectosylvian sulcus
ALs: Anterior lateral sulcus
Anss: Ansate sulcus
ARhs: Anterior rhinal sulcus
ASs: Anterior suprasylvian sulcus
Cors: Coronal sulcus
Crus: Cruciate sulcus
ELs: Ectolateral sulcus
Gens: Genual sulcus
M: Middle cerebral artery
MEs: Middle ectosylvian sulcus
MSs: Middle suprasylvian sulcus
P: Posterior cerebral artery
PCs: Postcruciate sulcus
PEs: Posteroir ectosylvian sulcus
PLs: Posterior lateral sulcus
PreSs: Presylvian sulcus
PRhs: Posterior rhinal sulcus
Pros: Prorean sulcus
PSps: Postsplenial sulcus
PSs: Posterior suprasylvian sulcus
Spls: Splenial sulcus
Ss: Sylvian sulcus
SSps: Suprasplenial sulcus

References

Bolk, L., E. Göppert, E. Kallius und W. Lubosch: Handbuch der vergleichenden Anatomie der Wirbeltiere. Bd. II/I, pp. 201—253. Berlin/Wien: Urban & Schwarzenberg 1934.

Campbell, A. W.: Histological studies on the localisation. of cerebral function. London: Cambridge University Press 1905.

Filimonoff, I. N.: Über die Varianten der Hirnfurchen des Hundes. J. Psychol. Neurol. (Lpz.) **36**, 22—43 (1928):

Górska, T.: Functional organization of cortical motor. areas in adult.dogs and puppies. Acta Neurobiol. Exp. **34**, 171—203 (1974).

Gurewitsch, M. und G. Bychowsky: Zur Architektonik der Hirnrinde des Hundes. J. Psychol. Neurol. (Lpz.) **35**, 283—300 (1928).

Hamuy, T. P., R. B. Bromiley and C. N. Woolsey: Somatic afferent areas I and II of dog's cerebral cortex. J. Neurophysiol. **19**, 458—499 (1956).

Hasegawa, S.: Morphological studies of the "cat" gyrus in the carnivora (dogs, foxes, cats) cerebrum. Anat. Report Hokkaido Univ. **143**, 1—51 (1959).

Herre, W. und H. Stephan: Zur postnatalen Morphogenese des Hirnes verschiedener Haushundrassen. Morph. Jb. **96**, 210—264 (1955).

Kawamura, K.: Variations of the cerebral sulci in the cat. Acta anat. (Basel) **80**, 204—221 (1971).

KLEMPIN, D.: Über die Architektonik der Großhirnrinde des Hundes. J. Psychol. Neurol. (Lpz.) **26**, 229–249 (1921).

KREINER, J.: Myeloarchitectonics of the central sulcus in the dog's brain. Bull. Acad. pol. Sci. Cl. 2. **9**, 481–484 (1961).

KUBOTA, S.: Comparative studies of the cerebral sulci on the lateral surface in dogs (shepherd, terrier, ainu dog). Anat. Report Hokkaido Univ. **33**, 1–62 (1956).

KUSAMA, T., K. OTANI and E. KAWANA: Projections of the motor, somatic sensory, auditory and visual cortices in cats. Progr. Brain Res. **21**A, 292–322 (1966).

MEYNERT, T. H.: Die Windungen der convexen Oberfläche des Vorderhirns bei Menschen, Affen und Raubtieren. Arch. Psychiat. Nervenkr. **7**, 1877. (cited from Hasegawa).

NAITO, J. and K. KAWAMURA: Variations of dog cerebral sulci. Acta Anat. (Nippon) **50**, 47–48 (1975).

NAKAHAMA, H.: Functional organization of somatic areas of the cerebral cortex. Int. Rev. Neurobiol. **3**, 187–250 (1961).

OTSUKA, R. und R. HASSLER: Über Aufbau und Gliederung der corticalen Sehsphäre bei der Katze. Arch. Psychiat. Nervenkr. **203**, 212–234 (1962).

RADINSKY, L.: Evolution of the canid brain. Brain Behav. Evol., **7**, 169–202 (1973).

SYCHOWA, B.: Degenerations of the medial geniculate body following ablations of various temporal regions in the dog. Acta Biol. exp. (Warszawa), **23**. 75–99 (1963).

TUNTURI, A. R.: Audio frequency localization in the acoustic cortex of the dog. Amer. J. Physiol. **141**, 397–403 (1944).

TUNTURI, A. R.: Further afferent connections to the acoustic cortex of the dog. Amer. J. Physiol. **144**, 389–394 (1945).

TUNTURI, A. R.: Physiological determination of the boundary of the acoustic area in the cerebral cortex of the dog. Amer. J. Physiol. **160**, 395–401 (1950).

WOOLSEY, C. N.: Organization of cortical auditory system: A review and a synthesis. In: Neural mechanisms of the auditory and vestibular systems. (G. L. Rasmussen and W. F. Windle, eds.), pp. 165–180. Springfield, Ill.: Thomas 1960.

YOSHIKAWA, T.: Nomina anatomica encephalica. Tokyo: Kanehara Shuppan 1962.

Prof. Koki KAWAMURA, M. D. Department of Anatomy, School of Medicine, Iwate Medical University, Morioka 020 (Japan)

Buchbesprechung

Paraneurons. New Concepts on Neuroendocrine Relatives. Proceedings of the Symposium on the Paraneurons, Morioka, Japan, April 1st, 1977.

Editors: S. Kobayashi and T. Chiba. Archivum Histologicum Japonicum, Vol. 40, Suppl. 1977.

Japan: Japan Society of Histological Documentation, 1977, 331 S., 171 Abb., 17 Tab.

Das Symposium, das als Satellitentreffen des 82. Japanischen Anatomischen Kongresses veranstaltet wurde, stellt eine Weiterführung des 1975 in Japan abgehaltenen Symposiums über chromaffine, enterochromaffine und verwandte Zellen dar, auf dem der Begriff der Paraneurone eingeführt wurde. Wie T. Fujita im einleitenden Referat feststellt, werden unter dieser Bezeichnung Zellen zusammengefaßt, die zwar nicht als Neurone anzusehen sind, mit diesen aber strukturelle, biochemische und embryologische Gemeinsamkeiten aufweisen, so daß eine exakte Abgrenzung nicht möglich ist. Paraneurone haben z. T. eine vorwiegend endokrine, z. T. eine sensorische Funktion, und einige sind als Interneurone zu bezeichnen. In 24 Vorträgen in englischer Sprache, die zum weitaus überwiegenden Teil von japanischen sowie von einigen nordamerikanischen, schwedischen und westdeutschen Forschergruppen gehalten wurden, werden dann unter physiologischen, biochemischen und strukturellen Aspekten die als Paraneurone bezeichneten chromaffinen Zellen des Nebennierenmarks, GABA-Zellen der Pankreasinseln, endokrinen Zellen des Gastro-entero-pankreatischen Systems, parafollikulären Zellen der Schilddrüse, Granula enthaltenden Zellen des autonomen Nervensystems, SIF-Zellen, Monoamin enthaltenden Geschmackszellen, Neuroepithelialkörperchen der Lunge, Hauptzellen des Karotidkörpers und Monoamin speichernden Zellen der Aorta und des Herzens, subependymalen Zellen des Hypothalamus, Pinealozyten und die neurosekretorischen Zellen des Nucl. paraventricularis bei verschiedenen Species abgehandelt. Das Buch schließt mit einer kritischen Einschätzung des vorangegangenen Symposiums über chromaffine, enterochromaffine und verwandte Zellen sowie des APUD-Konzepts von Pearse und des Konzepts der Paraneurone durch H. S. Bennett.

Der trotz des schnellen Erscheinens vorzüglich ausgestattete, insgesamt auf Hochglanzpapier gedruckte Band enthält eine Fülle neuer und neuester Versuchsergebnisse. Die Wiedergabe der zahlreichen elektronen- und lichtmikroskopischen Aufnahmen ist von guter Qualität. Alle Referate sind mit umfangreichen Literaturnachweisen versehen, die ein weitergehendes Studium einzelner Fragen erleichtern. Das Buch vermittelt somit einen ausgezeichneten Überblick zum aktuellen Wissensstand auf einem Gebiet, dessen Bearbeitung sich erst in den Anfängen befindet. Es stellt ein hochinteressantes Nachschlagewerk nicht nur für den Neurophysiologen, Morphologen, Biochemiker und Endokrinologen dar, sondern wird für alle von großem Wert sein, die sich mit den Grundlagen der Organisation und Regulation bei den Metazoen beschäftigen.

F. Döcke, Berlin

J. Hirnforsch. 19 (1978) 469–478

Laboratoire d'Histologie Embryologie et Cytogénétique de l'U. E. R. de Médecine et Biologie de Bobigny
(Université Paris XIII)

Effets cytologiques de divers fixateurs sur les neurones du ganglion de Gasser du foetus de rat

Par Claudine SAVY et Edmond TAMBOISE

Avec 15 figures et 2 tableaux

(Reçu Octobre 27, 1977)

Résumé: Cette étude est une comparaison de l'effet de plusieurs fixateurs sur la morphologie des neurones du ganglion de GASSER du foetus de rat de 18 jours, pour définir dans ce matériel le fixateur le plus adapté à la microscopie optique. Une description cytologique des neurones après coloration banale et une étude planimétrique permettent de dégager les qualités respectives de ces agents et de choisir les plus adaptés.

Zusammenfassung: In der vorliegenden Arbeit wurde die Wirkung verschiedener Fixantien auf die Morphologie von Neuronen des Ganglion GASSERI bei 18 Tage alten Rattenembryonen untersucht. Mit Hilfe der Lichtmikroskopie sollte die optimale Fixierung ermittelt werden. Die zytologische Darstellung der Neurone nach konventioneller Färbung und anschließende planimetrische Messungen erlauben die Bestimmung der verschiedenen Eigenschaften der verwendeten Agentien und die Auswahl der optimalen Variante.

Summary: The effects of several different fixatives on the morphology of the GASSER ganglion neurons in 18 days old foetal rats are compared, in order to determine the fixative most suitable for light microscopy work with this particular material. A cytological description of neurons after a conventional staining and a planimetric study permit to find out the respective qualities of these fixatives and permit therefore to choose the most appropriate.

Key words: foetus — fixatives — neurons — cytology — light microscopy.

Introduction

Toute étude cytologique morphologique, tant qualitative que quantitative, doit tenir compte de la qualité de la fixation. Ainsi pour les ganglions nerveux, rachidiens ou craniens, un problème de fixation s'est posé chez l'adulte: plusieurs auteurs ont pu lui attribuer, ainsi qu'aux traitements histologiques subséquents, l'origine des différents types de neurones observés (BACSICH et WYBURN 1953; PINEDA et al. 1967). Mais l'utilisation de techniques complémentaires (fixation par dessiccation, histochimie, autoradiographie, microscopie électronique) a confirmé leur existence réelle et a permis de décrire, au cours de la différenciation et de la maturation des cellules, les phénomènes biochimiques et morphologiques qui aboutissent à l'individualisation de deux types de neurones (KALINA et WOLMAN 1970; SARRAT 1970; GAIK 1973; SOBKOWICZ et al. 1973; LAWSON et al. 1974; PANNESE 1974). Toutefois, la difficulté qu'il y a à les distinguer chez les embryons (KALINA et WOLMAN 1970; GAIK 1973), nous a amenés à les considérer comme un type cellulaire univoque.

Dans l'étude de ces neurones en situations particulières, telles que croissance anormale ou perturbée expérimentalement qui fera l'objet de nos travaux ultérieurs, il nous est utile de disposer de données témoins, contrôlées et stables. C'est à ce titre que la fixation retient ici notre attention.

Sans faire une étude exhaustive de l'effet des fixateurs, nous proposons de rechercher le fixateur morphologique le plus adapté dans le cas d'un matériel précis, le ganglion de Gasser, à un moment précis, le 18ème jour de vie foetale. Parmi les fixateurs utilisés dans différentes études de morphogenèse du tissu nerveux en microscopie optique, ce qui est notre cas, nous pouvons noter certains des principaux retenus par les auteurs: le formol neutralisé ou non (KALINA et WOLMAN 1970; SOBKOWICZ et al. 1973; STAUDT et al. 1973); le formol salé (BISCONTE 1975); le Bouin (STAUDT et al. 1973; ROGERS et COWAN 1973); le Bouin alcoolique (DERER 1974); l'alcool éthylique (ANGULO 1951); le Carnoy (ROGERS et COWAN 1973; LAWSON et al. 1974); le Halmi (TAMBOISE et TAMBOISE 1966); le Zenker et le Zenker-formol (STAUDT et al. 1973). La plupart de ces fixateurs ainsi que quelques autres également cités (VONWILLER 1953) seront testés dans cette étude.

Materiel et methodes

1. Embryons. Pour cette expérience, des rattes Wistar (CNRS Gif) de 250 à 300 grammes, sont utilisées. Les frottis vaginaux sont faits systématiquement pour déterminer le jour de l'oestrus et assurer le contrôle hormonal. La présence de spermatozoïdes ou du bouchon vaginal vérifiant le coït, permet de définir le premier jour de la gestation. Au 18ème jour de la gestation, les rattes sont sacrifiées par élongation bulbaire. Les embryons sont prélevés et recueillis immédiatement en milieu isotonique, puis pesés avant fixation. Seuls les embryons de poids moyen sont retenus.

2. Fixateurs. Les fixateurs sont préparés selon Gabe (1968). Nous avons testé 7 groupes de fixateurs, à base d'alcool, de formol, d'acide acétique, d'acide picrique, de bichromate de potassium, de chlorure mercurique (sublimé) et de tétroxyde d'osmium. Leur composition est résumée dans le tableau 1.

Les temps de fixation et les différents lavages avant inclusion sont ceux prescrits. L'inclusion est faite dans la paraffine, au Technicon, dans des conditions soigneusement standardisées. Les coupes frontales à 5 μ, sont colorées par l'hématoxyline de Groat-érythrosine. Les neurones des ganglions de Gasser sont photographiés au même grossissement (90 × 10) avec un Ortholux-Orthomat de Leitz.

3. Planimétrie. Elle est faite selon la méthode de Benoit (1922). Les photographies des cellules sont agrandies au maximum et tirées sur papier dans des conditions strictement controlées et homogènes. Cent corps cellulaires pris au hasard sont étudiés dans chaque cas, mais à la condition que leurs noyaux paraissent coupés en leur centre et que leurs nucléoles soient présents. Les cellules, les noyaux et les nucléoles sont découpés et pesés. Les résultats sont exprimés en "poids papier" pour 100 unités.

4. Expression des résultats. En fonction des valeurs croissantes des paramètres choisis (taille des cellules, des cytoplasmes, des noyaux et des nucléoles), les quinze fixateurs sont classés pour chacun d'eux, dans l'ordre décroissant de leur effet "rétractant". Ces mesures sont faites en valeur relative puisqu'il est techniquement impossible de définir celles d'une cellule vivante.

Le rapport nucléo-cytoplasmique est établi pour chaque fixateur sur la somme de 100 noyaux et de 100 cytoplasmes. De même, les rapports nucléolo-cytoplasmiques représentent la mesure de 100 nucléoles sur celle de 100 cytoplasmes. Une valeur moyenne des rapports est calculée pour chaque cas à partir de la somme des rapports individuels.

Resultats

A. Résultats qualitatifs de l'action des fixateurs.

L'examen microscopique montre des cellules grossièrement arrondies et révèle des différences importantes, tant dans leur taille que dans leur aspect (figures de 1 à 15). Pour certains fixateurs la rétraction des

Tableau 1: La composition des fixateurs testés

produits de base → fixateurs ↓	eau	alcool	formol (1)	acide picrique	acide acétique	acide chromique	bichromate de potassium	chlorure mercurique	tétroxyde d'osmium	chloroforme	acide nitrique	acide tri-chloracétique	substances ajoutées (2)
Alcool absolu		+											
Méthanol-acétique		+			+								
Carnoy		+			+					+			
Alcool-formol-acétique	+	+	+		+								
Formol salé	+		+										NaCl
Regaud	+		+				+						$CaCl_2$
Bouin	+		+	+	+								
Bouin-Allen	+		+	+	+								
Bouin-Hollande	+		+	+	+								Urée acétate de cuivre
Duboscq-Brasil	+	+	+	+	+								
Halmi	+		+	+				+				+	NaCl
Zenker	+				+		+	+					Sulfate de Na
Zenker-formol	+		+				+	+					Sulfate de Na
Petrunkewitch	+	+			+			+			+		
Flemming	+				+	+			+				

(1) Le formol est neutralisé systématiquement par le carbonate de calcium.
(2) Substances ajoutées dans le but d'améliorer le résultat de la fixation (Martoja 1967).

structures cellulaires laisse de larges espaces vides entre les cellules: c'est le cas de l'alcool absolu, qui est certainement le plus traumatisant pour les cellules. A l'opposé, se situent le Zenker et le Flemming qui montrent des cellules très larges et très peu d'espace intercellulaire.

1. Cytoplasmes

Alors que le cytoplasme est nettement homogène et se colore de façon régulière après les fixateurs à base de formol, tels que le Regaud, le formol salé et le Zenker-formol, il le reste encore, bien que se colorant de façon beaucoup plus intense, après le Bouin et le Duboscq-Brasil. Le cytoplasme apparaît finement granuleux après fixation au Carnoy, au Bouin-Allen, au Zenker, à l'alcool-formol-acétique et au Petrunkewitch qui contiennent tous de l'acide acétique et l'on observe les granulations les plus grossières avec l'alcool absolu.

2. Noyaux

Le noyau parait habituellement volumineux et excentré, mais l'aspect de la chromatine, la taille et la répartition des chromocentres varient beaucoup d'un fixateur à l'autre. la chromatine est répartie de façon fine et régulière dans tout le noyau pour les fixateurs tels que le Zenker-formol, le Bouin-Allen, le Flemming, mais avec une légère densification à la périphérie du noyau comme pour le formol salé et le Regaud. La chromatine tend à s'agglomérer en mottes plus ou moins denses. Celles-ci sont réparties uniformément dans tout le noyau pour l'alcool-formol-acétique, le Halmi et le Bouin-Hollande. Enfin, ces amas paraissent se disposer sur des réseaux grossiers avec densification à la périphérie du noyau pour le Duboscq-Brasil, l'alcool absolu, le méthanol acétique et le Carnoy.

3. Nucléoles

Le nucléole enfin, varie considérablement en taille et en forme suivant le fixateur. Il semble d'autant mieux conservé que la chromatine nucléaire est plus fine. Il est alors gros et régulier. La chromatine associée y forme un liseré ou un amas parfois bien visible. C'est le cas du formol salé, du Regaud, du Zenker-formol. Il est plus irrégulier ou se perd dans les amas chromatiniens avec les autres mélanges fixateurs.

Tableau 2.: Les tailles (pour 100 cellules)

fixateurs		cellules		cytoplasmes		noyaux		nucléoles	
		vpp (1)	vr (2)	vpp	vr	vpp	vr	vpp	vr
Alcool absolu		36,47	1e	16,90	2e	19,57	1e	1,56	9e
Méthanol acétique		40,85	5e	16,74	1e	24,11	10e	1,67	11e
Carnoy		40,11	4e	18,74	6e	21,67	5e	1,53	8e
Alcool formol acétique		41,95	7e	18,84	7e	23,11	7e	1,44	4e
Formol salé		44,47	10e	20,15	9e	24,32	11e	1,70	12e
Regaud		38,56	2e	18,62	5e	19,93	2e	1,28	2e
Bouin		45,88	13e	21,01	11e	24,87	13e	1,80	15e
Bouin Allen		47,05	14e	23,90	14e	23,15	8e	1,66	10e
Bouin Hollande		40,85	5e	16,87	2e	23,97	9e	1,44	4e
Duboscq Brasil		44,99	12e	20,46	10e	24,53	12e	1,52	7e
Halmi		44,69	11e	19,26	8e	25,42	14e	1,43	3e
Zenker		42,82	8e	21,01	11e	21,82	6e	1,24	1e
Zenker formol		43,35	9e	22,16	13e	21,19	3e	1,71	13e
Petrunkewitch		39,09	3e	17,71	4e	21,39	4e	1,51	6e
Flemming		52,89	15e	25,02	15e	27,87	15e	1,71	13e
Ecart type	s	4,03		2,50		2,23		0,16	
Moyenne (3)	x	42,93		19,83		23,13		1,55	
Limites de confiance	LS	45,14		20,20		24,35		1,64	
	LI	40,72		18,46		21,91		1,46	
Coeff. variation	CV	7%		12,6%		9,6%		10,3%	

(1) vpp: valeurs poids papier.
(2) vp: valeur relative selon la classification indiquée plus haut
(3) moyenne affectée de son intervalle de confiance au risque 0,05

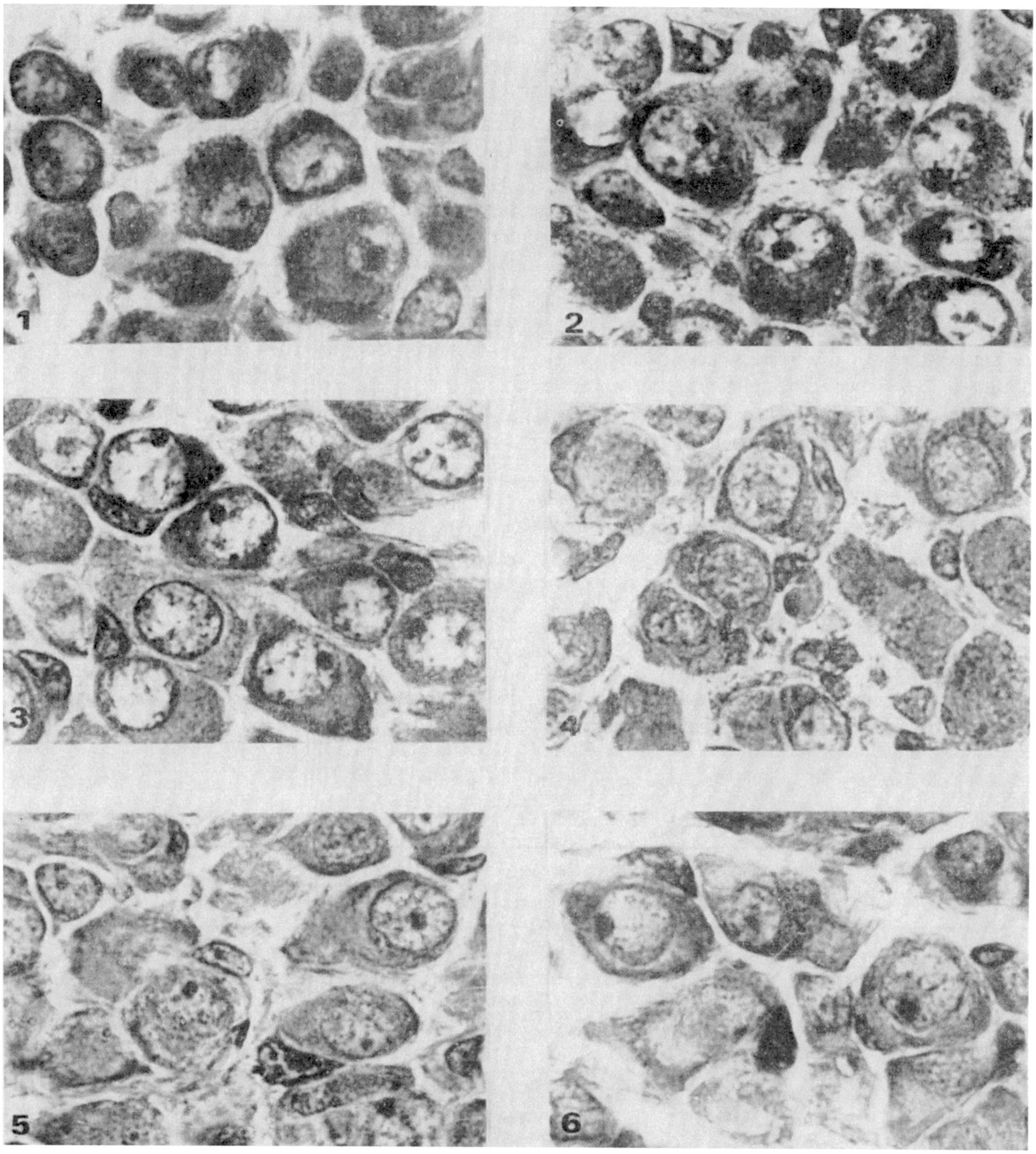

Fig. 1. Alcool absolu: cytoplasme à granulations grossières, noyau réduit, chromatine en réseau, nucléole relativement gros.

Fig. 2. Méthanol-acétique: cytoplasme granuleux, chromatine en réseau lâche, nucléole à contour irrégulier.

Fig. 3. Carnoy: cytoplasme finement granuleux, chromatine en amas disposés en réseau, gros nucléole.

Fig. 4. Petrunkewitch: cytoplasme finement granuleux, chromatine en fin réseau, nucléole peu visible.

Fig. 5. Alcool-formol-acétique: cytoplasme finement granuleux, chromatine en petits amas répartis dans tout le noyau, nucléole à contour irrégulier.

Fig. 6. Formol salé: cytoplasme homogène, chromatine légèrement densifiée à la périphérie du noyau. Liseré de chromatine associée, autour du nucléole.

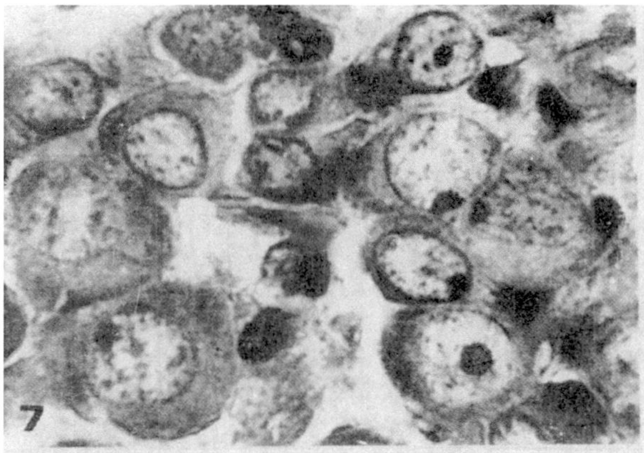

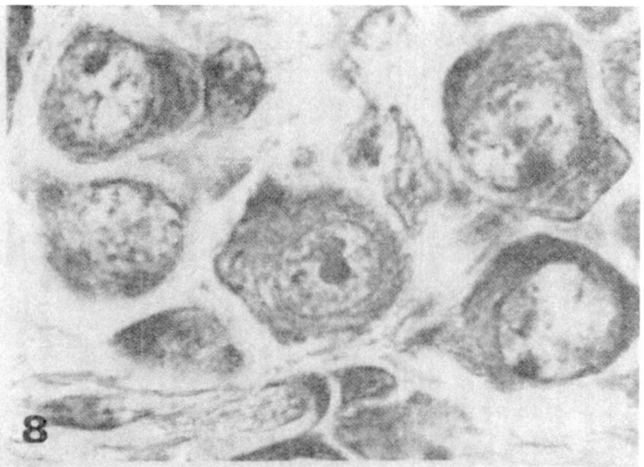

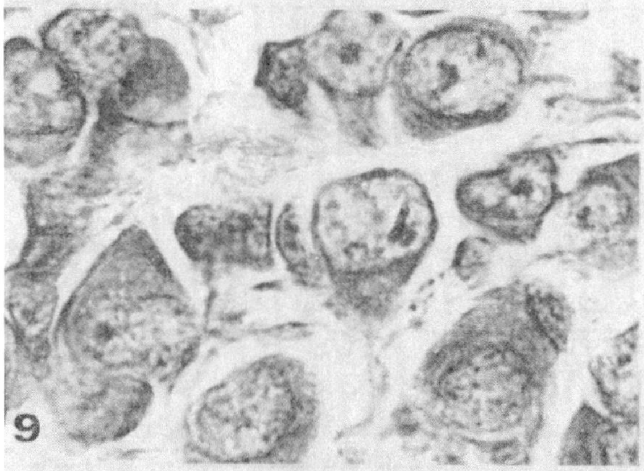

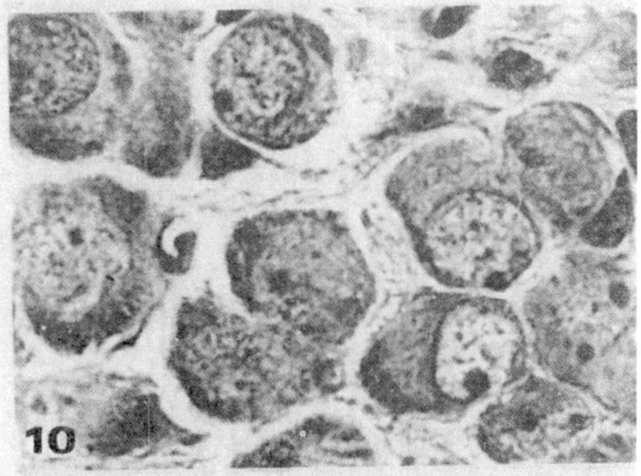

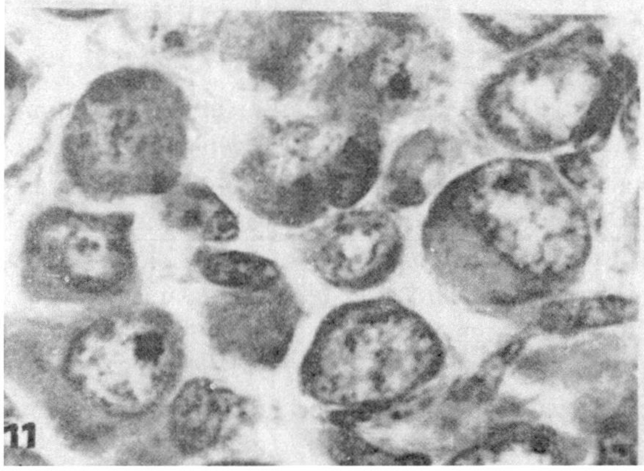

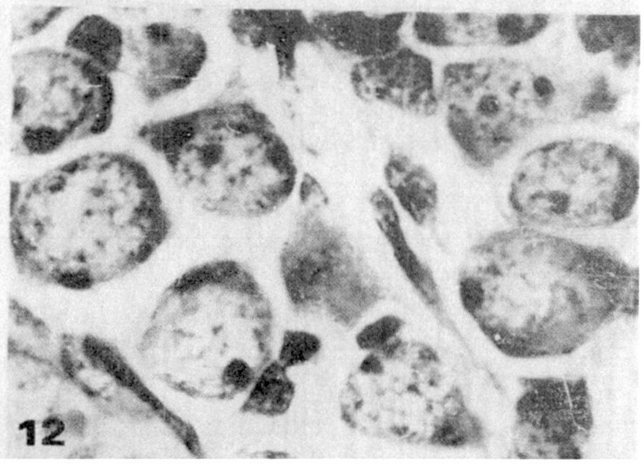

Fig. 7. Regaud: cytoplasme homogène, chromatine légèrement densifiée à la périphérie du noyau. Liseré de chromatine associée, autour du nucléole.

Fig. 8. Bouin: cytoplasme dense, chromatine finement réticulée, gros noyau à contour irrégulier.

Fig. 9. Bouin-Hollande: cytoplasme finement granuleux, chromatine en mottes plus ou moins denses, réparties dans tout le noyau, nucléole à contour irrégulier.

Fig. 10. Bouin-Allen: cytoplasme finement granuleux, chromatine fine et régulière, nucléole à contour irrégulier.

Fig. 11. Duboscq-Brasil: cytoplasme dense, chromatine en amas disposés en réseau, nucléole peu visible parmi les amas chromatiniens.

Fig. 12. Halmi: cytoplasme homogène et assez dense, chromatine en mottes plus ou moins denses, réparties dans tout le noyau, nucléole peu visible.

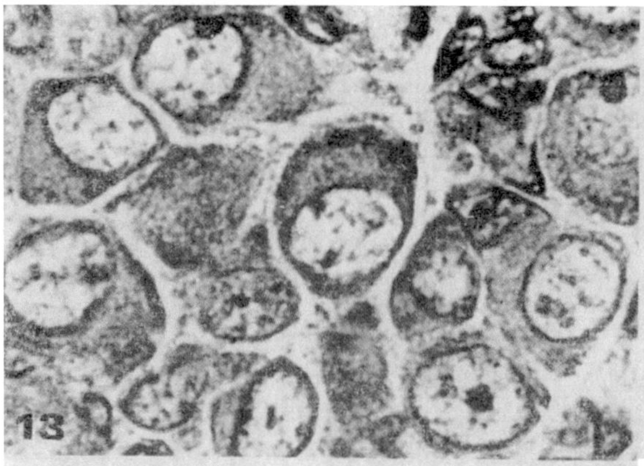

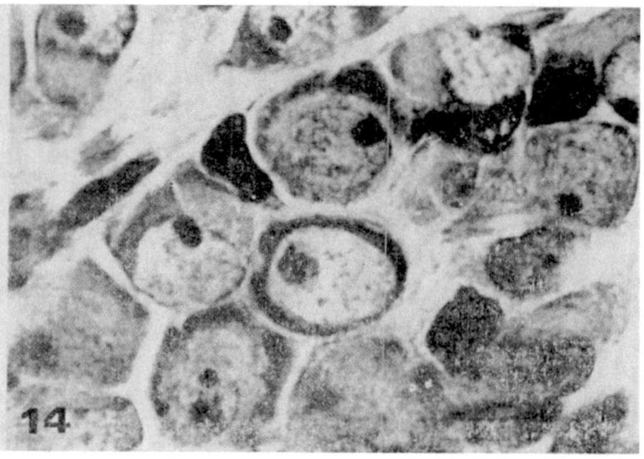

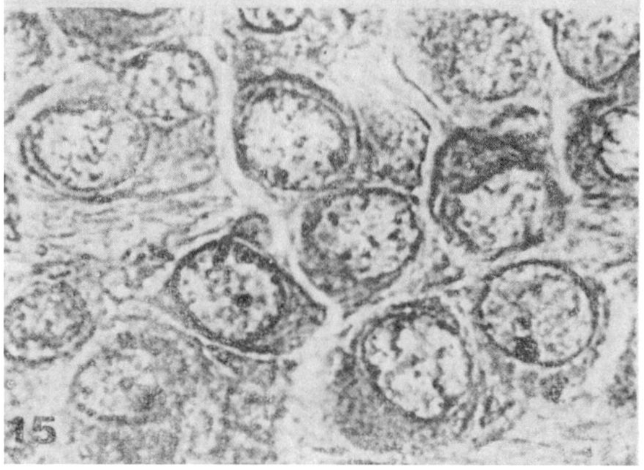

Fig. 13. Zenker: cytoplasme finement granuleux, chromatine finement réticulée, gros nucléole à contour irrégulier.

Fig. 14. Zenker-formol: cytoplasme homogène, chromatine fine et régulière, gros nucléole bordé par un liseré de chromatine associée.

Fig. 15. Flemming: cytoplasme finement granuleux, chromatine fine et régulière, nucléole peu visible parmi les amas chromatiniens.

B. Résultats quantitatifs de l'action des fixateurs sur la taille des cellules, des cytoplasmes, des noyaux et des nucléoles.

L'ensemble des résultats quantitatifs va dans le sens des observations morphologiques précédentes. Les paramètres mesurés pour les différents fixateurs sont rapportés au tableau 2. Ce sont les tailles cellulaires qui diffèrent le moins entre elles (cv 7%)*. Les variations des noyaux et des nucléoles sont du même ordre de grandeur (9,6% et 10,3%), alors que les tailles cytoplasmiques sont mettement plus différentes (12,6%). Il semble donc que ce soit le cytoplasme qui subisse avec le plus d'hétérogénéité l'action des fixateurs. Ce même tableau précise la classification globale en valeur relative.

La comparaison des rapports nucléo et nucléolo-cytoplasmiques (figures 16—17) met en évidence les modifications relatives dues à l'importance plus ou moins grande de l'action des fixateurs et leur impact différentiel sur les structures cellulaires (noyau, nucléole, cytoplasme). Ainsi, des fixateurs aussi différents que l'alcool absolu et le Flemming ont tous les deux, une action homogène sur le noyau et le cytoplasme (rapports nucléo-cytoplasmiques semblables et proches de la moyenne) mais tout à fait différente et hétérogène sur le nucléole et le cytoplasme (rapports nucléolo-cytoplasmiques très différents et éloignés de la moyenne); relativement au cytoplasme, le Flemming rétracte beaucoup plus le nucléole que l'alcool absolu.

Toutefois, l'examen de ces valeurs fait souvent apparaître des différences entre les fixateurs suivant la qualité du "produit de base" qu'ils renferment.

1. Fixateurs à base d'alcool

fixateurs	cellules	cyto-plasmes	noyaux	nucléoles
Alcool absolu	1e	2e	1e	9e
Petrunkewitch	3e	4e	4e	6e
Carnoy	4e	5e	5e	8e
Méthanol-acétique	5e	1e	10e	11e
Alc.-formol-acét.	7e	7e	7e	4e
Duboscq-Brasil	12e	10e	12e	7e

* Coefficient de variation.

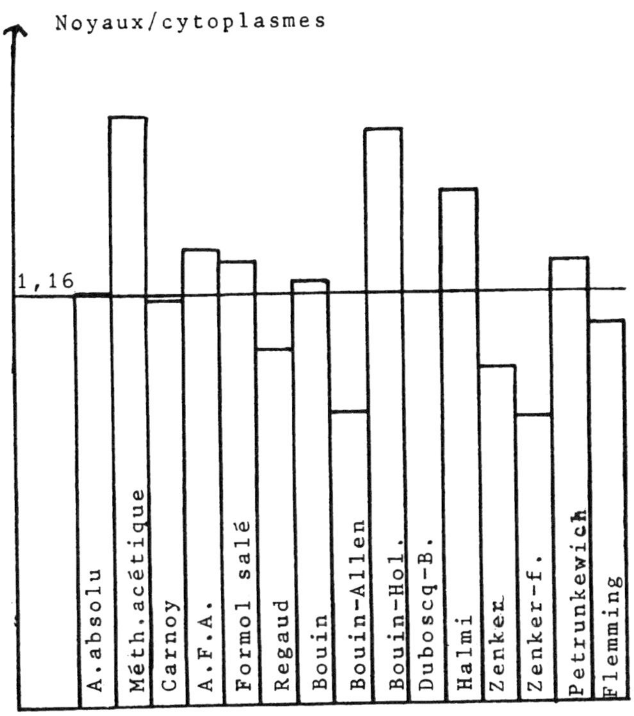

Fig.16. Rapport nucléo-cytoplasmique. Le rapport moyen est de 1,16.

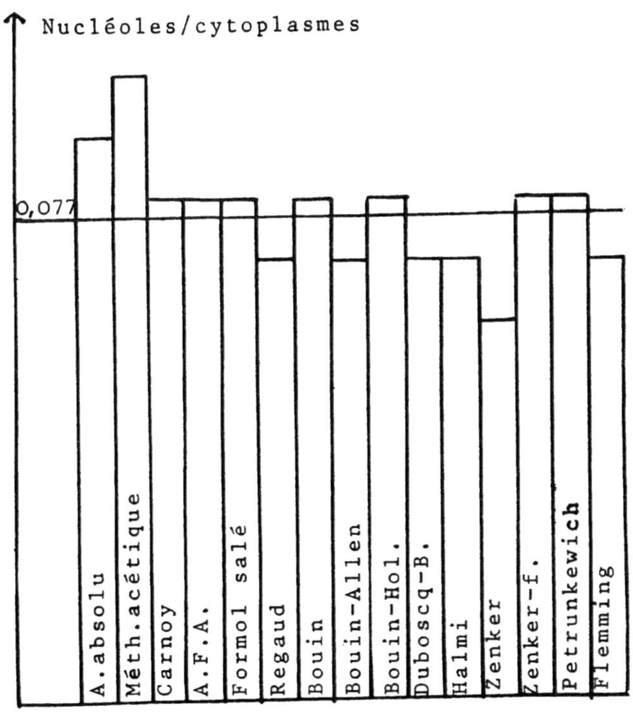

Fig.17. Rapport nucléolo-cytoplasmique. Le rapport moyen est de 0,077.

Bien que les tailles des cellules, noyaux et cytoplasmes se trouvent, ici, parmi les plus petites, on note cependant entre ces mesures, des différences importantes d'un fixateur à l'autre: ainsi, entre l'alcool absolu qui rétracte les structures au maximum et le Duboscq-Brasil (Bouin alcoolique) à l'opposé, se situe l'alcool-formol-acétique qui donne des valeurs moyennes. L'acide picrique du Duboscq-Brasil de même que le formol de l'alcool-formol-acétique interviennent favorablement.

Pour tous les fixateurs à l'alcool éthylique, le rapport nucléo-cytoplasmique est semblable, proche de 1,2; il évoque un effet homogène de la fixation sur les noyaux et les cytoplasmes. Il en va différemment du rapport nucléolo-cytoplasmique qui met en évidence un "comportement" particulier du nucléole face à la fixation alcoolique.

Le méthanol acétique a une position un peu particulière: la taille du noyau et du nucléole le situe dans les fixateurs les moins traumatisants, alors que le cytoplasme est au contraire très réduit.

2. Fixateurs à base d'acide picrique

fixateurs	cellules	cyto-plasmes	noyaux	nucléoles
Bouin-Hollande	5e	2e	9e	4e
Halmi	11e	8e	14e	3e
Duboscq-Brasil	12e	10e	12e	7e
Bouin	13e	11e	13e	15e
Bouin-Allen	14e	14e	8e	10e

Les tailles mesurées se situent en général parmi les plus importantes, l'acide picrique rétracte donc peu les structures cellulaires, le Bouin-Hollande faisant exception.

Ces fixateurs ont un effet homogène sur le nucléole et le cytoplasme (rapports nucléolo-cytoplasmiques semblables 0,07 et 0,08) alors qu'ils ont un effet hétérogène sur le noyau et le cytoplasme (rapports nucléo-cytoplasmiques de 0,97 à 1,42). Les fixateurs tels que Bouin, Bouin-Allén, Bouin-Hollande et Duboscq-Brasil ont une composition très voisine. Alors que le Bouin a une action très homogène sur les différentes structures cellulaires, une plus forte concentration en acide picrique (Bouin-Hollande) réduit très nettement leurs tailles. Seuls les noyaux et les nucléoles sont plus petits après adjonction d'urée (Bouin-Allen). Par contre avec le Bouin alcoolique (Duboscq-Brasil) la seule différence porte sur le nucléole dont la taille est réduite de 15% par rapport au Bouin.

3. Fixateurs à base de bichromate de potassium

fixateurs	cellules	cyto-plasmes	noyaux	nucléoles
Regaud	2e	5e	2e	2e
Zenker	8e	11e	6e	1e
Zenker-formol	9e	13e	3e	13e

Un caractère commun se dégage de ces résultats: les tailles nucléaires sont parmi les plus faibles alors que les valeurs cytoplasmiques se situent dans les plus importantes. Ceci se traduit par des rapports nucléo-cytoplasmiques minimaux et met en évidence un effet particulier de la fixation au bichromate de potassium sur les noyaux et les cytoplasmes. A noter que les tailles cytoplasmiques des cellules fixées au Regaud sont nettement inférieures à celles fixées par le Zenker ou le Zenker formol et qu'il existe entre ces trois fixateurs une dissociation importante des valeurs nucléolaires confirmées par les rapports nucléolo-cytoplasmiques.

4. Fixateurs à base de formol

fixateurs	cellules	cyto-plasmes	noyaux	nucléoles
Regaud	2e	5e	2e	2e
Bouin-Hollande	5e	2e	9e	4e
Alc.-formol-acét.	7e	7e	7e	4e
Zenker-formol	9e	13e	3e	13e
Formol salé	10e	9e	11e	12e
Halmi	11e	8e	14e	3e
Duboscq-Brasil	12e	10e	12e	7e
Bouin	13e	11e	13e	15e
Bouin Allen	14e	14e	8e	10e

Malgré leur diversité et à l'exception du Regaud et du Bouin Hollande, ces fixateurs se trouvent classés du 7e au 14e parce qu'ils provoquent une rétraction relativement faible du cytoplasme. En ce qui concerne les noyaux et les nucléoles, les résultats sont plus divers et difficilement comparables.

Plusieurs fixateurs montrent une très grande homogénéité d'action sur les cytoplasmes, noyaux et nucléoles: ce sont le Regaud, le formol salé et l'alcool-formol-acétique. Toutefois le Regaud occupe une position originale tant vis-à-vis des fixateurs formolés que de ceux à base de bichromate puisqu'il possède un pouvoir rétractant extrèmement important. Remarquons encore la position moyenne de l'alcool-formol-acétique dans cette classification.

5. Fixateurs à base de sublimé (chlorure mercurique)

fixateurs	cellules	cyto-plasmes	noyaux	nucléoles
Petrunkewitch	3e	4e	4e	6e
Zenker	8e	11e	6e	1e
Zenker-formol	9e	13e	3e	13e
Halmi	11e	8e	14e	3e

Pour cette catégorie de fixateurs les tailles cellulaires sont assez importantes bien qu'il y ait des différences entre eux: le Zenker, le Zenker-formol et le Halmi ont un effet comparable; le Petrunkewitch, très rétractant agit comme l'alcool, de façon forte et homogène.

Par contre le rapport nucléo-cytoplasmique ou nucléolo-cytoplasmique montre une grande hétérogénéité à l'exception du Petrunkewitch et du Zenker-formol pour le dernier. Il semble donc difficile de dégager un effet particulier du sublimé au point de vue quantitatif.

6. Fixateur à l'acide osmique: le Flemming.

Les cellules ainsi fixées montrent les tailles maximales. C'est aussi avec ce mélange qu'on observe les plus gros cytoplasmes, noyaux et nucléoles. Les rapports nucléo-cytoplasmique et nucléolo-cytoplasmique vont dans le même sens.

7. Fixateurs contenant de l'acide acétique

La plupart des fixateurs en contiennent en plus des autres substances déjà étudiées et de ce fait, il est difficile de dégager quelque chose de caractéristique sur le plan quantitatif.

Discussion

1. Qualité des cellules fixées

La forme des neurones ne nous parait pas affectée, en général, malgré les différences dues aux divers fixateurs. Les neurones du ganglion de Gasser des foetus de rat du 18ème jour, présentent tous un corps cellulaire arrondi ou relativement circulaire comme le définit le rapport entre grand et petit diamètre de la cellule, inférieur à deux selon Pannese et al. (1972) sur le ganglion spinal. Notre observation rejoint aussi celle de Lawson et al. (1974) faite en microscopie électronique sur le même matériel.

Selon ces auteurs les neurones spinaux du foetus, au 22e jour ont acquis cette forme. Ils commencent à ressembler aux neurones adultes, alors que ce n'est pas du tout le cas avant 15,5 jours de vie embryonnaire.

Par contre, nous ne retrouvons pas les cellules "multiangulaires" que BACSICH et WYBURN (1953) décrivent en microscopie optique dans le ganglion spinal adulte après fixation au formol salé, en plus des deux types de neurones connus. Ces éléments denses et homogènes fixent vigoureusement les colorants basiques. Pour ces auteurs, les neurones seraient "sensibilisés" par ce fixateur, aux traitements histologiques subséquents qui rétracteraient alors puissamment les cellules selon leur état fonctionnel. Toutefois, cet aspect particulier n'existerait pas quand le formol est associé à d'autres produits fixateurs dans un mélange différent. Cette interprétation paraît cependant discutable dans la mesure où les fixations à la glutaraldéhyde tamponnée-acide osmique du ganglion de Gasser du 18e jour font apparaître ce même type cellulaire sous forme d'éléments très colorés la paraphénylène-diamine. Examinés au contraste de phase, ils constituent des cellules denses présentant quelques fines vacuoles ou un aspect spongiocytoïde, avec un noyau irrégulier de type parapycnotique (TAMBOISE 1976). Il semblerait, en fait, que le formol salé puisse montrer des structures qui seraient détruites par les autres fixateurs.

2. *Qualité des noyaux fixés*

Nous observons des noyaux arrondis et ceci pour tous les fixateurs. LAWSON et al. (1974) situent l'acquisition de ce caractère aux jours qui précèdent la naissance du rat. De même, nous voyons souvent ces noyaux en position excentrée. Cet aspect est généralement considéré comme un caractère d'immaturité des neurones: GAIK (1973) le note dans le ganglion du nerf trijumeau d'embryon de poulet et SOBKOWICZ et al. (1973) dans le ganglion spinal du foetus de rat. La littérature cependant, n'apporte aucune précision sur l'effet des fixateurs dans la qualité du noyau et de sa position.

3. *Valeurs du quantitatif defini*

Le rapport nucléo-cytoplasmique est un bon moyen d'abord quantitatif puisqu'il fait intervenir la qualité de fixation du noyau et du catoplasme. Au 18ème jour, nous évaluons un rapport moyen de 1,16; le rapport minimal (0,96) est obtenu après fixation par le Zenker-formol et le maximal (1,44) après le méthanol-acétique. Le rapport moyen a donc une valeur élevée. Ce résultat va dans le sens des mesures de LAWSON et al. (1974) au 22ème jour qui indiquent dans ces neurones immatures une taille plus volumineuse du noyau par rapport au cytoplasme, que dans ceux du rat adulte.

Il y a ainsi des différences importantes d'un fixateur à l'autre. Pour ne pas masquer cet effet par une éventuelle réponse individuelle nous avons volontairement limité cette étude aux embryons de poids moyen. Aucun des auteurs cités n'évoque cet aspect du problème; ce qui peut paraître étonnant quand on admet qu'il puisse y avoir un lien entre la maturation des neurones, l'âge habituellement retenu et le développement des embryons.

Toutefois ,la comparaison des rapports nucléo-cytoplasmiques met en position particulière le méthanol-acétique, tout spécialement parmi les fixateurs alcooliques. Les neurones fixés par le méthanol-acétique ont des noyaux et des nucléoles très gros, dans des cytoplasmes très réduits, alors que les fixateurs à l'alcool éthylique provoquent une rétraction homogène des structures cellulaires. Le gonflement des tissus causé par le méthanol acétique est suivi par une rétraction due à la déshydratation et à l'inclusion (GERSH 1959). Nous constatons que celle du cytoplasme est très intense. Cet effet différentiel sur noyaux et nucléoles d'une part et cytoplasme d'autre part nous amène à penser qu'il est souhaitable d'éviter l'emploi de ce fixateur chez l'embryon plus sensible encore que l'adulte à cause de sa grande richesse en eau: il modifie et accentue la valeur du rapport nucléo-cytoplasmique des neurones immatures et introduit de ce fait, un risque d'erreur lors d'une étude quantitative comparée.

Conclusion

Il se dégage des résultats que l'alcool-formol-acétique, le formol salé, le Bouin et le Flemming sont les fixateurs les mieux adaptés à l'étude morphologique des neurones du ganglion foetal, parce que leur effet sur les structures cellulaires est le moins désorganisant et le plus homogène.

Ce choix répond à l'examen morphologique de cellules encore en différenciation. Il semble donc, qu'en plus des remarques faites couramment sur les problèmes de fixation du tissu nerveux, l'immaturité des neurones foetaux soit un élément qui justifie le soin apporté à la fixation dans tout travail qualitatif ou quantitatif sur leur croissance ou sa perturbation.

Bibliographie

ANGULO, A. W.: A comparison of the growth and differentiation of the trigeminal ganglia with the cervical spinal ganglia in albino rat embryos. J. Comp. Neurol. **95**, 53–71 (1951).

BACSICH, P., G. M. WYBURN: Formalin-sensitive cells in spinal ganglia. Quart J. micr. Sci. **94**, 89—92 (1953).

BENOÎT, J.: Sur une méthode permettant de mesurer la masse absolue du tissu interstitiel testiculaire. C. R. Soc. Biol. **87**, 1385—1387 (1922).

BISCONTE, J. C., R. MARTY: Analyse chrono-architectonique du cerveau de rat par radioautographie. I Histogénèse du télencéphale. J. Hirnforsch. **16** (1), 55—74 (1975)

DERER, P.: Histogénèse du néocortex du Rat albinos durant la période foetale et néonatale. J. Hirnforsch. **15** (1), 49—74 (1974).

GABE, M.: Techniques histologiques. Masson Ed. Paris (1968).

GAIK, G. C., A. I. FARBMAN: The chicken trigeminal ganglion. I An anatomical analysis of the neuron types in the adult. J. Morphol. **141** (1), 43—55 (1973).

GAIK, G. C., A. I. FARBMAN: The chicken trigeminal ganglion. II Fine structure of the neurons during development. J. Morphol. **141** (1), 57—75 (1973).

GERSH, I.: Fixation and Staining, In: The Cell, J. Brachet and A. E. Mirsky Ed., tome **1**, 21.

KALINA, M., M. WOLMAN: Correlative histochemical and morphological study on the maturation of sensory ganglion cells in the rat. Histochemie, **22**, 100—108 (1970)

LAWSON, S. N., K. W. T. CADDY, T. J. BISCOE: Development of rat dorsal root ganglion neurones. Studies of cell birthdays and changes in mean cell diameter. Cell Tiss. Res. **153**, 399—413 (1974).

MARTOJA, R., M. MARTOJA: Initiation aux techniques de l'histologie animale. Masson Ed. Paris (1967).

PANNESE, E., R. BIANCHI, B. CALLIGARIS, R. VENTURA, E. R. WEIBEL: Quantitative relationships between nerve and satellite cells in spinal ganglia. An electron microscopical study. 1. Mammals. Brain Res. **46**, 215—234 (1972).

PANNESE, E.: The histogenesis of the spinal ganglia. Adv. Anat. Embryol. Cell Biol. **47** (5), 1—97 (1974).

PINEDA, A., D. S. MAXWELL, L. KRUGER: The fine structure of neurons and satellite cells in the trigeminal ganglion of cat and monkey. Am. J. Anat. **121**, 461—488 (1967).

ROGERS, L. A., W. M. COWAN: The development of the mesencephalic nucleus of the trigeminal nerve in the chick. J. Comp. Neurol. **147** (3), 291—320 (1973).

SARRAT, R.: Zur chemodifferenzierung des Rückenmarks und der Spinalganglien der Ratte. Histochemie **24**, 202—213 (1970).

SOBKOWICZ, H. M., H. A. HARTMANN, R. MONZAIN, P. DESNOYERS: Growth, differentiation and ribonucleic acid content of the fetal rat spinal ganglion cells in culture. J. Comp. Neurol. **148**, 249—284 (1973).

STAUDT, J., P. STÜBER, G. DÖRNER: Morphologic studies of the matrix in the hypothalamus area in the rat depending on age and under experimental conditions. Z. Mikrosk. Anat. Forsch. **87** (5), 678—700 (1973).

TAMBOISE, E., A. TAMBOISE: Etude quantitative des modifications induites par les radiations ionisantes sur le ganglion de Gasser et ses cellules chez l'embryon de rat. Bull. Assoc. Anat. **51**, 979—988 (1966).

TAMBOISE, E.: Communication personelle (1976).

VONWILLER, P.: Nouvelles colorations nerveuses. C. R. Ass. Anat. **86**, 776—779 (1953).

Adresse des auteurs:

Laboratoire d'Histologie Embryologie et Cytogénétique
de l'U. E. R. de Médecine et Biologie de Bobigny
74, rue Marcel Cachin
93000 Bobigny

J. Hirnforsch. 19 (1978) 479–484

Catédra de Histología, Facultad de Biología, Universidad Complutense de Madrid

An ultrastructural study of the subependymal plate in the hamster

By

B. Fernandez Ruiz, I. Suarez Najera and A. Carrato Ibañez

With 7 figures

(Received October 20, 1977)

Summary: The subependymal plate of the hypothalamus of the hamster has been studied in this work. Our observations with the electron microscope show that the basal surface of the ependymal cells in this area are situated directly over the astrocytic elements. Ultrastructural studies which indicate two types of glial cells in the subependymal plate contradicted by the results of the present investigation demonstrating one type of glial cells in the hypothalamic subependymal plate. This fact is interpreted as the final result of a glial differentiation in adult hamsters.

Introduction

The ependymal cells which line the ventricular light are found separated from the neuropile by a stratum of variable thickness with elements of glial nature. With the optic microscope on the dog, Fischer (1967) suggests that said stratum is formed by astrocytes, microglia and two other types of subependymal cells. In the rat, Blakemore (1969) describes only one type of undifferentiated cells in the subependymal stratum or plate. Blakemore (1972), with the electron microscope, has also studied the components described by Fischer in the optic microscope, describing the microglial cells, and the two types of differentiated cells because of the nuclear aspect. In this work, we have attempted to observe the constitution of the subependymal plate in the hamster, with the electron microscope, limiting our study to the hypothalamic area.

Material and methods

The observations have made on adult hamsters, of both sexes. For the optic study the animals were anesthetized with eter, extracting the diencephalic block and introducing it into the Cajal's fixative, and following the silver carbonate method of Rio-Hortega (1942).

Sections of approximately 1 – 1'5 u thick were stained with toluidine-blue, mounted in Canada balsam and photographed in an optic microscope.

For the electron microscope study, the animals were fixed by perfusion with 2'5% glutaraldehyde (Milloning buffer, at pH 7'3). Later the ependymal and subependymal plates were isolated together and post-fixed in 1% osmium tetroxide. The blocks were dehydrated in acetone and embedded in Araldite. The thin sections were stained with lead citrate and examined in a Jeol 100-B electron microscope.

Results

A) Optic microscope

We have observed, with the silver carbonate method of Rio-Hortega (1942) a stratum of cells, the periependymal or subependymal cells, interposed between the ependymal cells and the neuropile (Fig. 1). These cells have a nucleus and cytoplasm clearer than the ependymal cells. In general, the periependymal cells have a thick process in perpendicular direction to the ependyma and the process is related to neurons or blood vessels.

The periependymal processes are larger in the upper part of the hypothalamus, in the area of the paraventricular and dorsomedial nuclei. In the figure 2 we can appreciate long astrocytic processes, running parallel to each other and, some of them, are parallel to the ependymal plate.

B) Electron microscope

Our observations with the electron microscope show us that the basal surface of the ependymal cells of the hypothalamic region are situated directly over the astrocytic somas or astrocytic processes.

These subependymal astrocytes are cells with a clear aspect and are similar to the ependymal cells. The nucleus is ovoid and presents invaginations, sometimes profund ones, as that observed in figure 5; the chromatina is disperse and present a pair of accumulations which are more dense at the margin of the internal nuclear membrane.

The astrocytic cytoplasm is not very dense at the electrons and has a moderate number of free ribo-

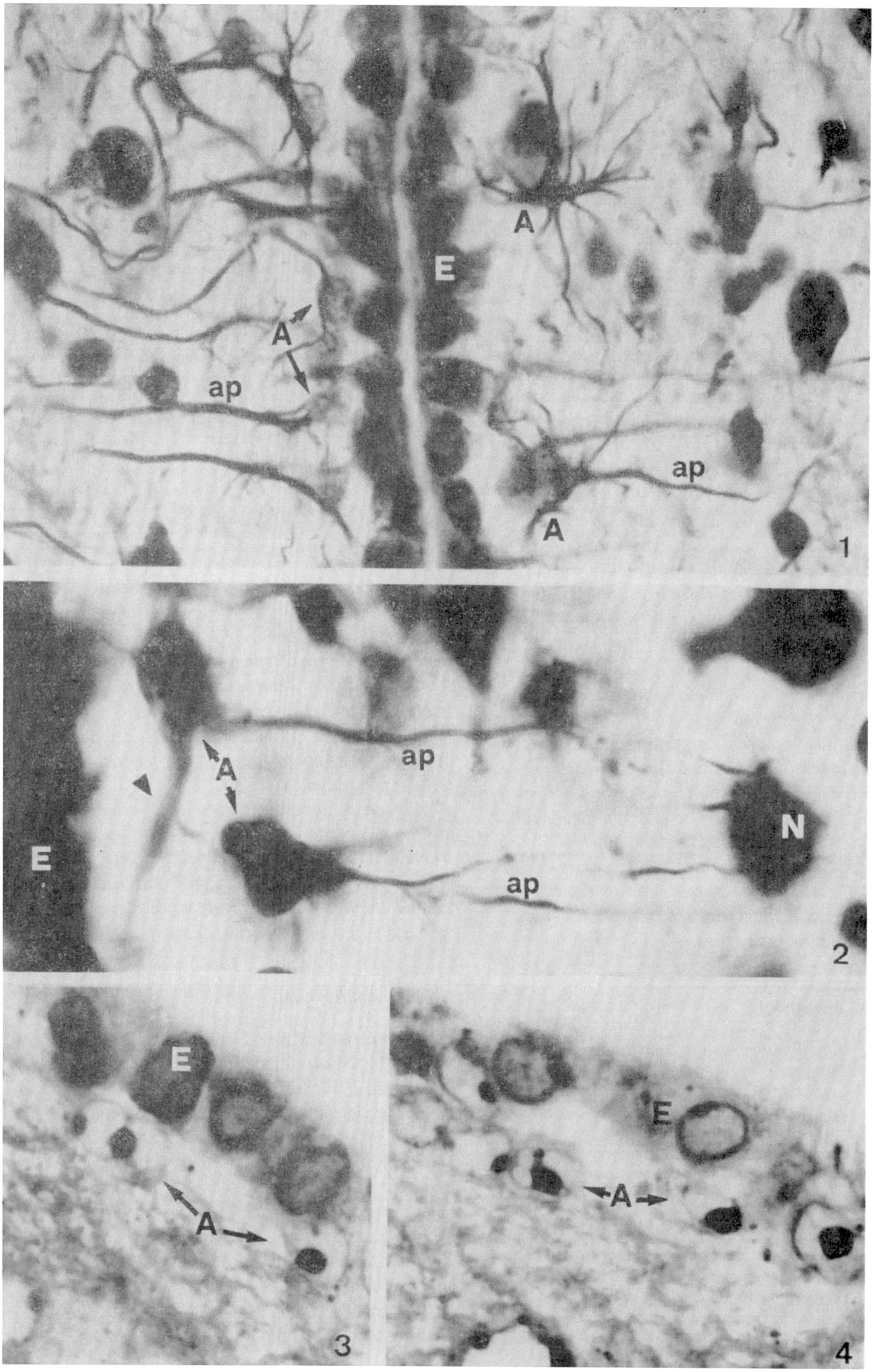
A
E
A
ap
ap
A
1
ap
A
ap
E
N
2
E
A
3
E
A
4

somes; its rough endoplasmic reticulum and the number of mitochondrias are scarce. The Golgi apparatus is well developed, with some of its vesicles quite dilated. The number of gliofibrils is particularly abundant (Figs. 6 and 7), situated between the cytoplasmic organules and extending in various directions. The plasmatic astrocytic membrane is located around the amyelinic nervous fibres gathering them into groups. The ependymal plasmatic membrane and the astrocytic membrane are so perfectly adapted that they leave no intercellular space between them. In general, the density of the ependymal cells is greater than that of the subependymal astrocytes.

The subependymal astrocytic processes run parallel to the surface of the ependyma. In figura 6 we observe the origin of an astrocytic process, whose diameter is practically occupied by a bundle of gliofibrils, whose fibrous components are slightly sinuous. Between the bundle of fibrils and the plasmatic astrocytic membrane we can see some free ribosomes. In the basal part of this process, we find an abundant number of amyelinic nervous fibres and in the apical part, another astrocytic process is interposed between the ependymal cell and the astrocyte to which we are refering. The two astrocytic processes run parallel, and with their base situated at the ependymal cell, are related to each other by means of intercellular junctions, of zonula adhaerens type. These junctions appear as thicknesses of the opposed cytoplasmic membranes, with greater electron density.

We have not observed vessels in contact with the ependymal cells but rather underneath the subependymal plate.

Discussion

If the morphology of the ventricular ependymal cells varies with the species, the age and their situation within the ventricular system (BLAKEMORE, 1972) the same be said for the subependymal plate.

←

Fig. 1. Parcial aspect of the hypothalamic ependyma (E) and the subependymal plate (A). The astrocytic processes (ap) run perpendicularly to the ventricular channal. Silver carbonate of RIO-HORTEGA (800×).

Fig. 2. Subependymal astrocytes (A) which processes (ap) are related to the hypothalamic neurons (N). Sometimes the subependymal processes are situated parallel to the ependymal cells (E) (triangle). Silver carbonate of RIO-HORTEGA (1660×).

Fig. 3 and 4. Semithin sections of 1,5 μ thick, showing the ependymal cells (E) and the subependymal plate (A). The astrocytes are clearer than the ependymal cells. Toluidine-blue (1660×).

Our observations are closer to the observations in the rat (BLAKEMORE, 1969) than in the dog (BLAKEMORE, 1972). In relation to the observed in the rat, we agree that there is one cellular type, not indifferentiated, but they are authentic fibrous astrocytes. We have not observed the different types of cells with a clear nucleus and cells with a dark nucleus that have been described in the dog. Blakemore's proposal to group all of the subependymal component under the name of the "E.T.A. Series" (ependymocites, tanycites and astrocytes) is not valid in the case of the hamster. The differentiation of cells with light and dark nucleus is based by the cited author on the possibility of the first representing ectopic localizations of the primitive germinative layer, while the latter would be the covering layer. LEWIS (1968b) suggests the possibility that these cells which immigrated to the covering layer would be converted into the adult brains in glia. The animals studied by us have all been adults and the cellular type identified is definitely astroglia.

BLAKEMORE (1972) prefers to include the cells that he calls of dark nucleus as the authentic cells of the subependymal plate, arguing that they are less differentiated and more similar to the spongioblasts of immature animals (MELLER and Co., 1966). To admit these cells as exclusive components of the subependymal plate has been confirmed by many toher authors: GLOBUS and KUHLENBECK (1944); BRYANS (1959); SMART (1961); LEWIS (1968) and CAVANAGH and LEWIS (1969).

We think that the difference of appreciation between these authors and ourselves can be motivated by the age of the animals and by the topographic region. The third ventricle divides the hypothalamus in two halves and this is a region in which abundant nuclei exist situated near the ependyma: suprachiasmatic, paraventricular, ventromedial, dorsomedial and infundibular, which justifies the necessary present of astrocytic elements among the ependimocytes and the neuronal components. The fact that we have only observed one cellular type is interpreted by us as the final result of a differentiation which perhaps was initiated in the fetal period and directed to the formation of astrocytes as the only component of the subependymal plate of the adult hamster.

We have already pointed out in the results that we have not observed blood vessels in relation to the ependymal cells which limit the third ventricle, which contrasts with that observed by BLINZINGER (1962) in the hamster, and in the rat (BRIGHTMAN, 1965; HIRANO and ZIMMERMAN, 1967, and WESTERGAARD, 1970). Our observations situate the capillaries underneath the subependymal plate.

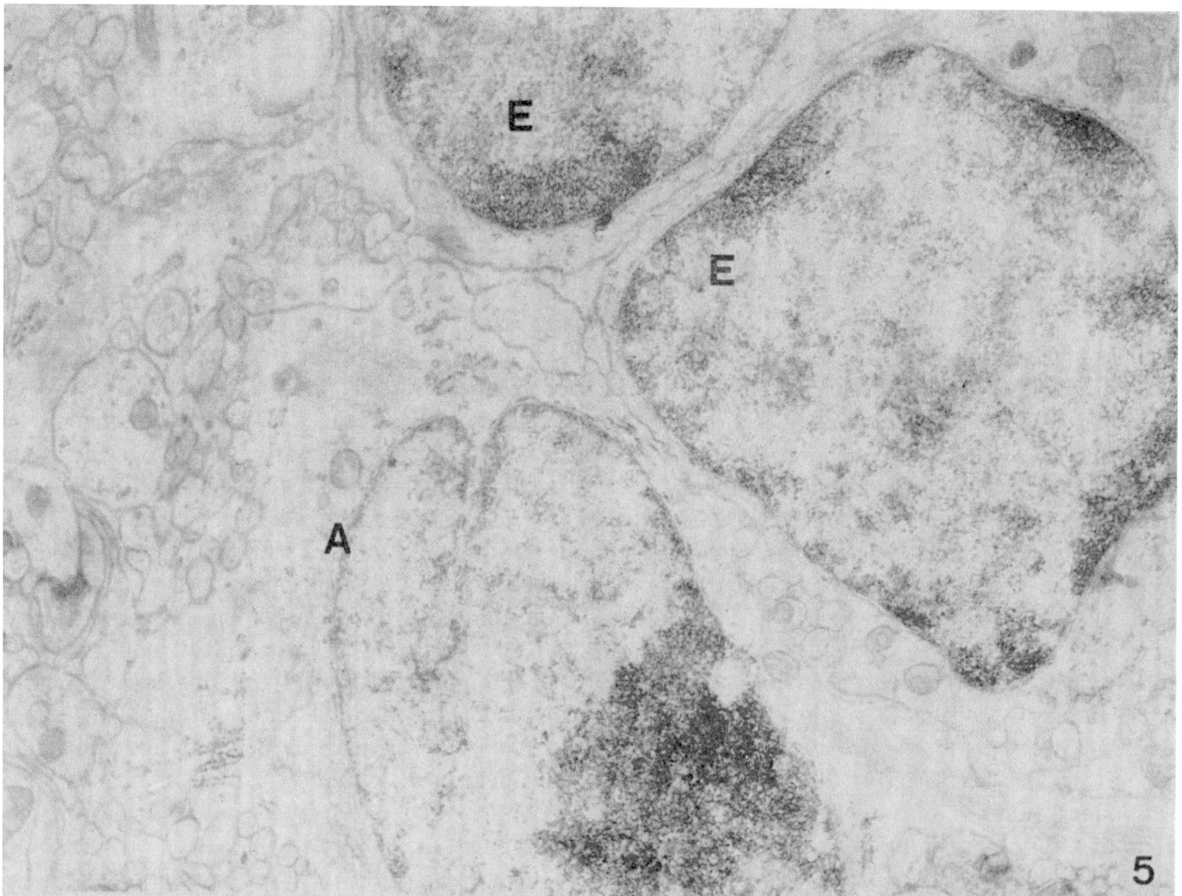
E
E
A
5

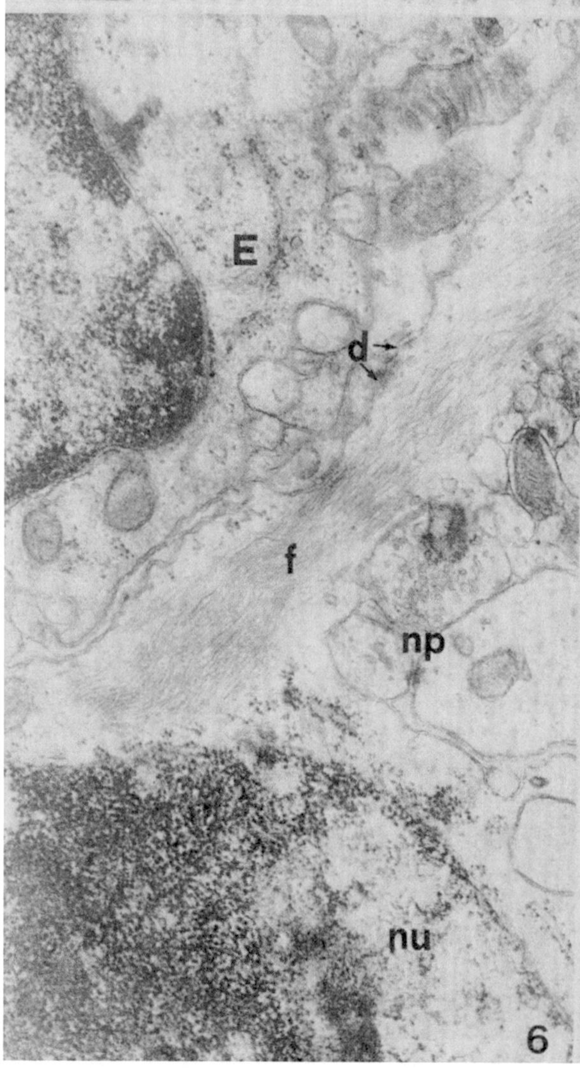
E
d
f
np
nu
6

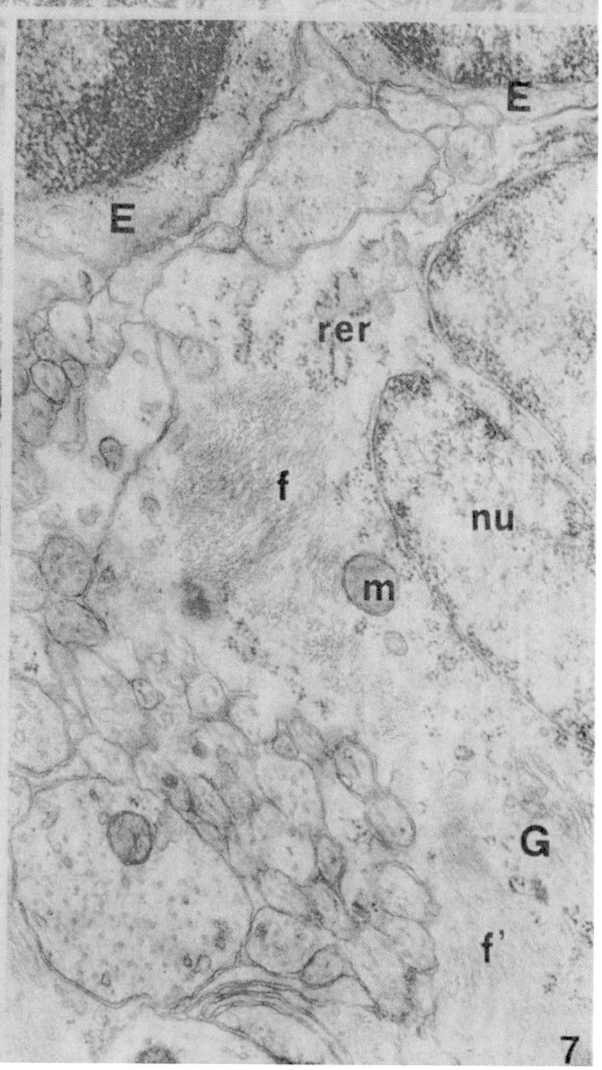
E
E
rer
f
nu
m
G
f'
7

Thus, the study of the hypothalamic subependymal plate in the hamster has shown that no basal membrane exists between the ependymal cells and the neurophile, but what does exist is an astrocytic stratum on which they lean and by means of which they are related to the nervous parenchima.

References

BLAKEMORE, W. F.: The ultrastructure of the subependymal plate in the rat. J. of Anatomy (London), **104**, 423—433 (1969).

BLAKEMORE, W. F. and R. D. JOLLY: The subependymal plate and associated ependyma in the dog. An ultrastructural study. J. of Neurocytol., **1**, 69—84 (1972).

BLINZINGER, K.: Elektronenmikroskopische Untersuchungen am Ependym der Hirnventrikle des goldhamsters (Mesocricetus auratus). Acta Neuropathol., I, fasc. **5**, 527—532 (1962).

BRIGTHMAN, M. W.: The distribution within the brain of ferritin injected into cerebrospinal fluid compartments. I) Ependymal distribution. J. Cell Biol. **26**, 99—124 (1965).

BRYANS, W. A.: Mitotic activity in the brain of the adul rat. Anat. Rec., 133, 65—74 (1959).

CAVANAGH, J. B. and P. D. LEWIS: Perfusion fixation, colchicine and mitotic activity in the adult rat brain. J. of Anatomy (London), 104, 334—350 (1969).

FISCHER, K.: Subependymale Zellproliferationen und Tumor-disposition brachycephaler Hunderassen. Acta Neuropathol., (Berlin), **8**, 242—254 (1967).

GLOBUS, J. H. and H. KUHLENBECK: The subependymal plate (matrix) and its relationship to brain tumours of the ependymal type. J. of Neuropathol. and Exp. Neurology, **3**, 1—35 (1944).

HIRANO, A. and H. M. ZIMMERMANN: Some new cytological observations on the normal rat ependymal cell. Anat. Rec., **158**, 293—302 (1967).

LEWIS, P. D.: Mitotic activity in the primate subependymal layer and the genesis of gliomas. Nature (London) **217**, 974—975 (1968, a).

LEWIS, P. D.: The fate of the subependymal cell in the adult brain, with a note on the origin of microglia. Brain, **91**, 721—736 (1968, b).

MELLER, K., W. BREIPOHL and P. GLEES: Early cytological differentiation in the cerebral hemisphere of mice. Z. Zellforsch., **72**, 525—533 (1966).

RIO-HORTEGA, P. del: El método del carbonato argéntico. Revisión general de sus técnicas y aplicaciones en histología normal y patológica. Arch. Histol. Nor. y Patol., **I** and **II**, 165—205 (1942).

SMART, I.: The subependymal layer of the mouse brain and its cell production as shown by autoradiography after thymidine-H^3 injection. J. Comp. Neurol., **116**, 325—347 (1961).

WESTERGAARD, E.: The lateral cerebral ventricles and the ventricular walls. M. D. Thesis, University of Aarhus (1970)

Author's address:

B. Fernandez RUIZ
I. Suarez NAJERA
A. Carrato IBAÑEZ
Catédra de Histología
Facultad de Biología
Universidad Complutense de Madrid
Madrid - 3, Spain

←

Fig. 5. Subependymal astrocyte (A) located basally to two ependymal cells (E) (16.800×).

Fig. 6. Gliofibrilar bundle (f) of the periependymal astrocyte, which fibrils run parallel to the ependymal cell (E). The astrocytic process is situated between the ependymal cell (E) and a group of nervous processes (np). (d: desmosomes) (27.600×).

Fig. 7. Astrocytic cytoplasm full of gliofibrils (f) in transversal section, near the rough endoplasmic reticulum (rer) and a mitochondria (m). G: GOLGI complex, near another gliofibrilar bundle (f') (27.300×).

Buchbesprechung

S. N. OLENEV: Razvivajuščijsja mozg. Kletočnye, molekuljarnye i genetičeskie aspekty nejroėmbriologii.
Leningrad ‚Nauka', 1978.

(Das Gehirn in der Entwicklung. Zelluläre, molekulare und genetische Aspekte der Neuroembryologie.) 222 Seiten mit 52 Abbildungen und 2 Tabellen. 2,40 Rbl.

Das vorliegende Buch ist der Entwicklung des Nervensystems auf der zellulären, molekularen und genetischen Ebene bei Hühnern und Ratten gewidmet. Es basiert auf einer umfangreichen Literaturkenntnis sowie eigenen Untersuchungen des Autors.

Der Text ist derart gegliedert, daß mit einer Betrachtung der embryonalen Quellen des Nervensystems begonnen wird. Anschließend erfolgt die Beschreibung der allgemeinen Gesetzmäßigkeiten der Phylo- und Ontogenese des Nervensystems anhand konkreter Beispiele: spinale und vegetative Ganglien, Rückenmark, Medulla oblongata, Kleinhirn sowie Mittel-, Zwischen- und Endhirn. Gleichzeitig wird die chronologische Reifung dieser Abschnitte bei Hühnern und Ratten aufgezeigt.

Weiterhin werden Angaben gemacht zur morpho-funktionellen Organisation einander entsprechender Abschnitte des Gehirns bei Wirbeltieren. Zum Schluß der Betrachtung werden Grundhypothesen erörtert, die die Mechanismen der Entwicklung des Nervensystems klären helfen. Die verschiedenen obengenannten Abschnitte des Nervensystems werden unter den angegebenen Gesichtspunkten jeweils für Hühner und Ratten abgehandelt, so daß bei insgesamt 222 Seiten einem Teilgebiet nur 1 bis 2 Seiten gewidmet sind, in denen einem auf diesem Gebiet arbeitenden Wissenschaftler kaum neue Aspekte mitgeteilt werden können.

Das Buch ist als eine Art Lehrbuch zur Einführung in das Gebiet von Entwicklung und Bau des Nervensystems denjenigen Wissenschaftlern zu empfehlen, die mit der Arbeit am Nervensystem erst beginnen.

Die 52 Abbildungen sind in der Mehrzahl schematische oder halbschematische Zeichnungen, die zum Teil schwer zu erfassen sind (zu klein und dadurch unübersichtlich).

Das umfangreiche Literaturverzeichnis mit 415 zitierten Arbeiten läßt u. a. Autoren wie BRODMANN, CAJAL, KIRSCHE, LORENTE DE NÓ, MARIN-PADILLA, ROSE, SARKISSOW, STEPHAN, VOGT vermissen. Als Anhang ist ein Sachverzeichnis enthalten. Besonders hervorzuheben ist der günstige Preis von 2,40 Rubel.

B. SCHÖNHEIT, Berlin

Kurzinformationen für Autoren

1. Publikationsgebiet

Das Journal für Hirnforschung veröffentlicht Arbeiten der experimentellen, descriptiven und vergleichenden Neuroanatomie, Neurohistologie, Neuroembryologie auf der Grundlage licht- und elektronenmikroskopischer, histochemischer Untersuchungen und unter Anwendung der Methode der Zell- und Gewebezüchtung. Es werden auch Arbeiten multidisziplinären Charakters aufgenommen, sofern sie morphologische Ergebnisse beinhalten.

2. Allgemeine Bedingungen

Zur Veröffentlichung angenommene Arbeiten dürfen in keiner anderen Zeitschrift angeboten werden. Die Publikation folgender Arbeiten ist möglich:

1. Originalarbeiten nicht länger als 30 Schreibmaschinenseiten.
2. Kurzbeiträge nicht mehr als 2 Schreibmaschinenseiten.
3. Übersichtsartikel mit kritischer Interpretation bedeutender Forschungsergebnisse.
4. Arbeiten zur Geschichte der Neuroanatomie, Neurohistologie und Neuroembryologie sowie Nachrufe.
5. Buchbesprechungen.

3. Die Arbeiten können in deutscher, englischer und französischer Sprache veröffentlicht werden.

4. Manuskripte

Die Manuskripte müssen sorgfältig geschrieben sein (Schreibmaschine, doppelter Zeilenabstand).

4.1. Die erste Seite enthält:
- 4.1.1. Die Institution aus der die Arbeit hervorgegangen ist und in der Regel Name des Direktors.
- 4.1.2. Titel der Arbeit.
- 4.1.3. Vorname(n) und Name(n) des(r) Autors(en).
- 4.1.4. Bei Arbeiten in deutscher und französischer Sprache ist eine englische Zusammenfassung einschließlich einer englischen Übersetzung des Titels der Arbeit beizufügen.
- 4.1.5. Kurztitel von nicht mehr als 50 Buchstaben.

4.2. Originalarbeiten sollten folgende Teile enthalten:
- 4.2.1. Einleitung und Problemstellung.
- 4.2.2. Material und Methode.
- 4.2.3. Ergebnisse.
- 4.2.4. Diskussion.
- 4.2.5. Literatur.
- 4.2.6. Vollständige Anschrift des oder der Autoren.

5. Literatur

Der Autor ist verantwortlich für die Übereinstimmung der Zitate im Text mit dem Literaturverzeichnis. Alle im Literaturverzeichnis enthaltenen Beiträge müssen im Text erwähnt sein. Im Text ist anzugeben: Name des Autors und in Klammern Jahr der Publikation. Bei Beiträgen mit mehr als 3 Autoren wird nur der Erstautor genannt mit dem Vermerk et al.
Das Literaturverzeichnis enthält: Namen sämtlicher Autoren, Vornamen abgekürzt, Titel der Arbeit, Name der Zeitschrift, Band, Anfangs- und Endseitenzahl, Jahreszahl in Klammern.

6. Abbildungen

Die Abbildungen sind fortlaufend als Textfiguren zu numerieren. Die Bildunterschriften müssen dem Manuskript beigefügt werden. Von allen Abbildungen mit eingezeichneten Buchstaben usw. muß eine zweite unbeschriftete Vorlage beiliegen. Hinsichtlich der Abbildungsgröße ist der Satzspiegel von 17,6 × 25,7 cm zu beachten.

7. Versand

Alle Arbeiten sind sorgfältig verpackt eingeschrieben zu versenden.

Manuskripte nehmen entgegen:

Prof. Dr. J. Anthony

Museum National d'Histoire Naturelle, Paris V/France, 55, Rue de Buffon

Prof. Dr. A. Hopf

Institut für Hirnforschung der Universität Düsseldorf, D - 4000 Düsseldorf, Moorenstr. 5, Gebäude 22.03.05

Prof. Dr. W. Kirsche

Anatomisches Institut der Humboldt-Universität zu Berlin, DDR - 104 Berlin, Philippstr. 12

Prof. Dr. J. Szentágothai

Anatomisches Institut der Universität Budapest, Budapest (Hungary) Tüzoltó utca 58

Inhalt

Contents

Indexed in Current Contents, Philadelphia

FSC
www.fsc.org
MIX
Papier | Fördert gute Waldnutzung
FSC® C083411

Zeitfracht Medien GmbH
Ferdinand-Jühlke-Straße 7
99095 Erfurt, Deutschland
produktsicherheit@kolibri360.de